# STEM CENTURY

# STEM CENTURY

## It Takes a Village to Raise a 21st Century Graduate

## 21STCENTED

# Contents

# Introduction

Education was organized for work, work has changed, education has not.

Over time, the purpose of education has shifted according to the needs of the country. The primary purpose of education has ranged from instructing our children in religious doctrine, to preparing them to live in a democratic society, to assimilating immigrants into the mainstream of society, to preparing workers for the industrialized 20th-century workplace.[1]

A poll conducted by PDK/Gallup on education summarized the following:

> Parents and adults (each 53%) agree that academics should be the focus of a public school education while teachers say the main goal should be preparing students to be good citizens (45%) or preparing them academically (37%)....Another question, though, makes clear that learning job skills is a broadly shared

goal nonetheless: Seventy-five percent of parents, 77% of all adults, and 82% of teachers say schools should prepare students both for jobs and academically. Only 15% of parents and all adults and 8% of teachers say core academics alone should be the priority.[2]

---

Since the dawn of mankind, education has been about survival of the species. From the time of the hunter-gatherer to farming to the industrial revolution, our fundamental purpose for education was to survive and thrive. The fundamental skills we used were critical thinking, creativity, collaboration, and communication. We went from practice to theory in making our world more predictable. Whether confronting the threat of the saber-tooth tiger or determining where to hunt and gather, we used critical thinking, creativity, collaboration, and communication—skills that educators today call the 4Cs.

Today, our saber-tooth tiger is digitization, and deciding where to hunt needs STEM education. As we did in our early days, we need to go from practice to theory so that our young people can survive and thrive in the rest of the 21st century.

Beginning with the homo sapiens, we have been inventing the tools needed for our survival. Technology has always been, and will always be, at the heart of our inventions. Every generation used the technology of their time to advance their society. We use technology for fire, tools, weapons, the wheel, the printing press, the internet, and now mass digitization.

We continue to manipulate our technology in such a way as to eliminate our physical and mental labor for many jobs, and soon most jobs. Universities, businesses, and other thought leaders across the globe continue to study the impact of technology on the modern-day workforce. The Oxford University Study on Employment indicates that 47 percent of all jobs will be eliminated or impacted by technology.[3] The McKinsey study suggests that 50 percent of all jobs are automatable and 30 percent will be automated by 2030.[4] The World Economic Forum reports that 6.1 million jobs will be created because of automation.[5] Early in the pandemic, my prediction was that COVID-19 would impact these numbers as employers look to guard against another year like 2020 by automating as many functions as possible using current and emerging technologies.

For the rest of the 21st century, we will accelerate our investments in STEM fields. Robotics in manufacturing, self-driving trucks in transportation, AI-reading X-rays in medicine, robots making pizzas in fast food, and drones making deliveries in retail are just the tip of the iceberg of what we will see. The results of this acceleration will force us to move beyond students with the natural disposition for STEM to attract all students to STEM. We must ensure a comprehensive STEM foundation for ALL students, just as we do with numeracy and literacy. The confluence of technology, such as artificial intelligence, automation, robotics, and other emerging technologies, is transforming our world in a way that will leave us well-educated but useless without a foundation in STEM education.

## A 21st-Century Education

Today's education has been organized by the Prussians around the late 1700s for the Industrial Age. It was brought to the United States by Horace Mann in the mid-1800s and has stayed the same since. Education was then standardized to offer credentials so people could qualify for work. Features like

grouping by age, with a specific amount of time to gain knowledge, then pacing through the system whether they are ready
or not are dubbed "the factory model of education."

Alvin Toffler, who ardently denounced the "Industrial Era
School" in his book *Future Shock,* wrote:

> Mass education was the ingenious machine
> constructed by industrialism to produce the kind of
> adults it needed. The problem was inordinately
> complex. How to pre-adapt children for a new world
> —a world of repetitive indoor toil, smoke, noise,
> machines, crowded living conditions, collective disci-
> pline, a world in which time was to be regulated not
> by the cycle of sun and moon, but by the factory
> whistle and the clock.

The solution was an educational system that, in its very structure, simulated this new world. This system did not emerge
instantly. Even today it retains throw-back elements from pre-
industrial society. Yet the whole idea of assembling masses of
students (raw material) to be processed by teachers (workers) in
a centrally located school (factory) was a stroke of industrial
genius. The whole administrative hierarchy of education, as it
grew up, followed the model of industrial bureaucracy. The
very organization of knowledge into permanent disciplines
was grounded on industrial assumptions. Children marched
from place to place and sat in assigned stations. Bells rang to
announce changes of time.[6]

Arguably, education was organized for work, work has
changed, but education has not. We have been talking about

the transformation of education for many years with little noticeable change. The COVID-19 pandemic created urgency to change a system that has been too big to change. Schools failed, people stopped working, and industry came to a standstill. The aftermath is that industry has leveraged technology to evolve itself to accommodate its own transformation and thus will require a STEM-ready workforce.

Here again industry will dictate how to educate the workforce of the future. The pandemic resulted in remote work, cloud-based collaboration, video conferencing, and worker-centric accommodations allowing for kids to run around in the background shattering "the work–life balance." As such, the acceleration of technological breakthroughs occurred, and life will never be the same.

Advancement in AI, automation, and robotics has caused, and will continue to cause, the replacement of physical labor, eliminating jobs in every industry. As education did before, it will evolve to accommodate the needs of industry. Industry moved away from its own factory-based model, and now education must do the same. But how? The answer is a comprehensive STEM education.

STEM—science, technology, engineering, and math—when taken together as an integrated discipline, provides opportunities for students to develop the enduring skills of critical thinking, creativity, communication, and collaboration. A STEM project, by its nature, is an experience in the skills dubbed as 21st-century skills or the 4Cs. So, back to survival.

For the 21st century and beyond, education must be about skill and talent development, which is also workforce development, and the economic vitality of communities. Unless we learned a specific skill or studied a profession, most of us can't remember what we learned in school. This isn't a bad thing because assimilated knowledge makes for a well-rounded person; however, project-based learning with a focus on mastery is a time-tested method of education that is needed again in today's schools.

Mastery-based education is using practice to validate principles and theories, and it is the way education sticks. Social emotional learning deals with the whole child, which should focus on not just how they learn but also what they learn. So, interest-based learning is crucial. Self-paced learning is a must for full-potential learning. Hybrid learning, blended learning, and remote learning provide crucial options where location-based learning is a challenge for students and their families. The number of families home-schooling their children has never been higher. With a flexible learning environment, this number would drop significantly.

## Workforce Development

We need all students to be prepared for the workforce. We simply will not have enough people to fill the demand for workers without them. There are just over 625,000 computer science jobs available, with only 71,000 computer science students graduating into the workforce last year.[7] We will need 2.4 million advanced manufacturing workers,[8] 2.7 million data scientists,[9] and 465,000 cybersecurity professionals[10] in the United States over the next few years. To complicate matters, the gaps we have mentioned are not in the United States alone; the STEM workforce shortage is a global problem.

Major global economies, such as China, India, and countries in Europe, are faced with the same STEM workforce challenges as in the United States. While the projected cybersecurity shortage in the United States is just under 500,000 jobs, in India, for example, some are estimating that there will be a cybersecurity job gap of 1.5 million.[11] Developing countries are preparing their young people to go after these jobs with a fierce urgency.

While the United States pays some of the most competitive wages and can keep its workers, some of these developing nations are better preparing their young people to take jobs from the United States without leaving their native land. Still, the threat to a community is not just from a better prepared international community. Within the United States, there is a brain drain happening in local communities as they face competition from other regions across the nation. It's imperative, now more than ever, that schools take, again, their rightful place in workforce development and economic development.

Given all of this, *It Takes a Village to Raise a 21st-Century Graduate*. The school district must be the center of transformation. They have the future workforce and the resources to make a systemic change. It is not easy and will not be easy until the outdated education policies are changed to support work that needs to be done. We cannot wait for policies to change or we will lose generations! Instead, let's heed the old African proverb and reach out to the village to raise the child.

If we are going to leverage the village to raise the child, which we must, then the role of the school superintendent must be different. The school superintendent is the most important role in a community—they have our children! They have the awesome responsibility of shaping young minds, developing responsible citizens, and training a workforce for the continued economic vitality of this great nation. Yet, when conversations about workforce development and economic development are being had, they are often not at the table.

Superintendents of schools must assume their rightful place among the village elders, expand the fours wall of the district, and not ceremoniously but meaningfully include the stakeholders of the community. The community intelligence must inform instruction and create opportunities not otherwise available in the insular world of education. They must create partnerships that provide access to resources bilaterally. Curriculum resources should be shared with the community, and the community should provide real-world learning opportunities for students.

I am excited to have the opportunity to share the stories, perspectives, and vision for STEM education from superintendents across the country. The rest of the 21st century is the most crucial period in our lifetime, and a pivotal century for mankind.

My mission through 21stCentEd[12] is to make STEM education available and accessible to all students, in school, after school, and out of school. Working with schools, nonprofits, and communities, we are making sure that students engage and grow in STEM and are provided the tools to reach their full potential. We are equally, if not more, determined to ensure underserved black and brown students, girls, and rural communities get the support they need to participate confidently in what is coined by the World Economic Forum as the Fourth Industrial Revolution.

The United States is being challenged from all sides for our innovative dominance. India, China, and others are investing heavily to ensure that they become globally competitive and catch us while we sleep. As such, the STEM Century is about education, workforce development, and economic prosperity so that America can continue to lead in innovations, technology, and economic vibrancy. The people at the heart of these innovations will determine the direction of our society while making a great living and changing the trajectory of life for their families for generations to come.

Marlon Lindsay
Founder | CEO

# About the Author

MARLON LINDSAY

Marlon is a business leader and writer who is committed to helping people reach their full potential. He believes that the stewardship of our young is a biological imperative and thus our ultimate priority. He pursues this priority daily with his own six children and the work that he does at 21stCentEd.

He creates STEM opportunities for students early to prepare and capitalize on the disruptions brought on by artificial intelligence, automation, robotics, and other emerging technologies. He works towards the imperative of getting students ready for a STEM-centric economy through public/private partnerships in education, workforce readiness, and the economic development of 21st-century communities.

Since the COVID-19 pandemic, he has doubled down on ensuring that *all* students, especially the underserved and underrepresented, have a foundation in STEM by bringing affordable and scalable comprehensive STEM education to school districts and their communities.

He is motivated to influence a world where his children can play, create, collaborate, and reach their full potential. Marlon is noted for saying, "The best way I know to make sure my kids reach their full potential is to ensure that all kids reach their full potential—adults are just big kids."

Born in Jamaica, Marlon immigrated to the United States at the formative age of thirteen. He grew up in Connecticut, where he earned his undergraduate and master's degrees from the University of Connecticut—Go Huskies! He furthered his learning, training in leadership, financial management, and business development, at Wharton and Columbia, Dartmouth respectively. He is the author of *Reminder to Self, The Eight Life Principles for Living from Your Truest Self.*

He lives in Utah with his wife, Amanda, where they are proud parents of Skye, Makenna, Bliss, Shiner, River, and Wilder.

ONE

# The Need for STEM with Latino Students

Elizabeth Alvarez, Ph.D.

In 1984, as an eighth grade Latina, I stood in line outside my grammar school entrance door. I had no interest in entering the building. I knew my questions and curiosity would not be welcomed in my school. First, because I was Latina, and second, because the teachers in this institution did not value questioning, productive struggle, and innovation in education. As I stood outside in line in my brown plaid uniform in my knee-high socks, I looked up at the sky. It was close to seven o'clock in the morning and the moon was as clear as it would be in the darkest night—a beautiful, round white moon with its familiar marks of shadow.

I am sure there were other days besides this one that allowed me to connect with science, but this day is significant. I could not understand nor was ever taught by the teachers in this school why I could see the moon when it was morning. As the bell rang, I could not take the image of the moon out of my head. On this same day, my "science teacher" told me I was failing science. I had asked her what I needed to do to graduate. She said it was a long shot but perhaps teaching the class, the subject that I was failing, for two weeks. The irony. It was these words, which showed lack of support, confidence, and reassurance that began my first taste and love of teaching science. If the moon can come out in the daylight, then I could teach the course that I was failing.

Here is what I know about science. Science relies on other disciplines. I also know that in science there is no right or wrong; there is, rather, proven or unproven. Therefore, I knew either it was proven that I did not know science, and that is why I was failing, or unproven and there was another reason I was failing.

I became a science middle school teacher in 1992, and I was the only person of color at that middle school and the only middle school teacher who also spoke Spanish. The moon appearing during the day left me with questions, wonder, and curiosity that one teacher tried to stump or kill within me at the young age of twelve but instead sparked a fire to pursue a subject that I had thought was not meant for me. When speaking about science, there is a relationship between math and technology. Together, they produce the scientific endeavor and what makes their unity successful.

Here is what I know about math. Math relies on both logic and creativity. It relies on numbers that produce trends and patterns, looks for correlations and relationships. Therefore, scientists and engineers apply math to their own work and use math to solve problems. It is also known that the ancient Aztec empire invented math.

Here is what I know about technology. Because of the beautiful relationship between science and math, technology was produced. It is through engineering that complex systems and tools have been developed. It is through technology that science and math realize that there are endless possibilities. It is through technology that the Aztec empire built a large and powerful island of Tenochtitlan.

Here is what I know about Latinos in STEM. There is a small percentage of Latinos who enter STEM careers, although opportunities in the workforce for STEM are growing every year. Based on the *Hispanic & STEM* report in collaboration with the Research Consortium on STEM Pathways, "Hispanic adults, historically underrepresented in STEM, continue to be an underutilized talent pool. They currently comprise 17% of the workforce overall but merely 8% of the STEM workforce."[1]

When I was in grammar and high school, I never had a Latino educator. This is significant. I also was exposed to only two science courses in high school. This was not an error but a choice made by others who were not my family. Latinos complete fewer STEM coursework in high school, making it difficult for them as they enter college to choose STEM as a career. This lack of exposure is one of the reasons why Latinos do not choose to move into STEM fields.

It was during my journey in getting my Ph.D. that I became immersed with my student's stories in science. As a middle school science teacher, I learned the importance of dialogue and the power of storytelling. This was not something I learned in graduate school but from my Latino students. When we are born, we are born with so many questions, and as young children, we are constantly inquiring. We utilize the "why, how and what," and for many children, this is not encouraged but seen to be disrespectful. These questions are not followed with, "Let's find out?" or "I don't know, but I bet if we look into it or test it out or do a little experiment, we will find out."

We also dismiss their stories as not being relevant or connected to the world around us. Talking science means asking those questions. It means sharing our stories of what we do know so we can compare our stories and begin to see patterns and comprehend. I insisted that my students come with their stories and talk it through. How is it that a liquid can become a solid with heat or cold? My children brought their stories of their mom's cooking pancakes and how heat solidifies the batter to make a pancake. They would share taking water and putting it in an ice tray to make ice. These stories, along with allowing them to show me, began their love of the STEM world. STEM is dialogue and doing, two concepts that at times dissipate in particular schools as our children get older. This love of dialogue and doing must be cared for from a very young age. Exposure to STEM begins in the early years.

In my experience in leadership, if you go into any US school, it is unlikely that you find even one Latino teaching science. Children see this and begin to believe that this is not a subject or career for them when there is no representation in the school for the subject. I was always told by principals that I was a unicorn because not only was I Latina but I also spoke Spanish to support our emergent bilingual students. These children were normally self-contained and not departmental-ized because there was no bilingual Spanish-speaking science teacher. Many schools do not expose our emergent bilingual students with an endorsed science teacher. No wonder there is an achievement gap in STEM for Latinos.

There is also the ruling hand that occurs for our Latino students, the policing that occurs as they grow older, so schools keep order and management at the forefront rather than teaching and learning. Questioning is seen as defiant. I have seen this firsthand as a teacher. Students were good students if they sat quietly in their seats finishing their worksheets where they can find answers in a book rather than by talking among themselves. They are not to have agency, authority, or identity for themselves, particularly in science. Accomplishing work quietly in some subjects are easy, but STEM becomes a challenge because it cannot be taught without the back and forth of language, making mistakes, asking questions, and making calculations. STEM is a social activity.

As a principal, I was determined to make sure my children would be exposed to science at an early age. I also knew that STEM was in our blood. Our Mayan ancestry was the first in many scientific discoveries between about 300 and 900 AD. We excelled in astronomy, agriculture, engineering, and communications. The Aztecs built Tenochtitlan. Talk about engineers building a remarkable island city! They are known to have developed mathematics, their own calendar, and contributions to medicine. Our children as young as five were entering science fairs and knew the scientific method, a method that I use to this day as a leader. By the time they reached middle school, my Latino students were competing every year in competitions and winning. This exposure allowed them to pursue further coursework in high school feeling prepared and ready to conquer further post-secondary coursework. If children are not encouraged or exposed early on, which is the case for many Latino students, they begin to doubt their interest, their passion, and their capabilities, although it is in their blood.

There is an equity issue when it comes to how we are exposing our students. It starts with the lack of Latino science/STEM educators as well as preparation and connection to STEM, the resources needed for our schools with BIPOC students, the importance to see bias within assessments, and the lack of family and school connection, even more when there is a language barrier. There is a need to allow dialogue and kinesthetics for STEM coursework for Latinos.

As a superintendent, I already see the need to support the good work. I currently have one Latino teacher and two Black teachers in my middle schools. None of them teach STEM. I have no Latino K–5 teachers and only two Black K-5 teachers. There are other factors that I have not discussed yet. One that I see to be a challenge is the support for social emotional awareness that our BIPOC students receive versus their white counterparts. It is important to address this, as I see the need in my district. It is most evident because of the pandemic, but it has always been there. The well-being of our children and what occurs at home as well as the relationship children have with their educators contribute to the confidence level Latino children feel with STEM. This is a definite need where social emotional curriculum should be embedded within the curriculum. The need for teachers to begin to understand and build relationships allows students to begin to love learning.

The day I saw the moon during the day and learned I was failing science was the day I knew I would be a science teacher. I taught two weeks the way I wanted to be taught. The way I needed to be taught. I allowed my classmates to touch objects and question. I wanted them to have two-way conversations where they pushed each other's thinking. I engulfed myself into reading research so that I could share this with them and allow them to compare, connect, and understand. I graduated with a D in science, but I knew all along that science was in my blood. From that day, I was determined to make sure other children like me would pursue their questions and use Science, Technology, Engineering, and Math to make sense of the world around them, such as the moon being out during the day. The Mayans and Aztecs exposed their children at an early age and understood the relationship they would have with STEM and what could and would be created and developed.

I raised two children with my love of STEM, and they too fell in love with it; both entered the medical field. It was not pushed or forced upon them; rather, their curiosity was encouraged and nurtured. I have seen it with my own students, both as their teacher and as their principal. By allowing them to make mistakes, go back to the drawing board, and do it all over until it was proven or not proven, the love begins to grow and further questions arise. Let us follow in our ancestors' footsteps and allow our Latino students to rediscover their connection and relationship to STEM. We will see them thrive because, like me, it is in their blood.

# About the Author

## ELIZABETH ALVAREZ, PH.D.

Dr. Elizabeth Alvarez was born and raised in the Southwest Side of Chicago. She is a first-generation Mexican and first-generation college graduate. She received her B.A. at the University of Illinois at Chicago in Education and minored in Psychology. She continued with her education in completing two masters and a Ph.D. through the University of Illinois at Chicago. She received her Ph.D. in curriculum and instruction that focused on narrative, race, and science pedagogy. The title of her dissertation is Rabbit on the Moon: An Urban Mexican Curriculum Story. Liz began her career as a teacher in the Little Village Neighborhood in Chicago. She taught for thirteen years teaching upper grade science. She was privileged and honored to serve as principal for six years at John C. Dore Elementary, a pre-K through eighth grade elementary school in Chicago's Clearing neighborhood where she led the school to level 1+ status.

Prior to being an administrator, she coached district principals in math and science, and was an adjunct professor at Concordia University. She is active with the Network of Hispanic Administrators in Education for Chicago Public Schools, a member of the Illinois Association of Latino Administrators and Superintendents and current president of

IALAS, 2019 cohort of the Superintendent Leadership Academy of ALAS, a 2015 Fellow of Leadership Greater Chicago and Corwin reviewer for book proposals. She is a member of the CPS DREAMERS Fundraiser supporting DACA students achieve their college dreams, and she is one of the founders and board member of the Latino Leadership Pipeline that coaches and supports Latino leaders as they work to become future leaders in education. Her past role was serving as Network 8 Chief of Schools in Chicago Public Schools supporting eighteen schools (predominantly Latino) where she moved schools from the 30 percent to the 70 percent where they achieved growth and attainment along with preparing students for college and career readiness.

She was awarded the Chicago Mayoral Award for Level 1 principal 2012 and 2013, Independent School Principal Recognition 2015 by Chicago's mayor, Chicago's top 23 Latina Leaders in Philanthropy and Government for 2017 and 2018, and the 2018 Latina Leader Award by Chicago Latino Network.

She is currently the proud superintendent of Forest Park District 91. She is the first female, Latina, and person of color to lead the district and one of four Latina superintendents in the state of Illinois. However, her proudest title is mother of two great children, Natalia and Caleb. As a leader, she encourages staff, community, and students to achieve greatness in all they do and take pride in their school climate and culture. More importantly, she aims to instill a sense of belonging with a developed understanding of social emotional learning driven to make schools safe learning environments so that learning is at the forefront and increased student achievement is inevitable.

TWO

# STEM Programs versus a STEM Education

### Dr. Maria Armstrong

As a former teacher, administrator, superintendent, and now as the Executive Director for the Association of Latino Administrators and Superintendents, a national organization, let me share with you one of the top three areas that keep me up at night: preparing our youth for a tomorrow that doesn't yet exist and what to do for the students who are in need of skillsets for the careers and professions of today. My career began in private industry. You might, for all intents and purposes, consider me a late bloomer to education, as I did not enter the education system by traditional means. What I learned while working alongside most teachers is that they typically went to college and, for a whole host of reasons, graduated college and immediately became teachers.

I worked in the electronics field for nearly a dozen years before becoming a teacher. It was from that experience of "adapt or seek other employment" that helped me understand the importance of research and development and the funding that allows for industries to stay nimble, innovative, and profitable. This experience also made it second nature for me to see the science, technology, engineering, and math within the curriculum standards. It was fun for me to have students learn about science, tech, engineering, and math (STEM) without having to say, "Today, at this hour, we will be learning math." Instead, I was able to braid STEM subject matters into each day by utilizing everything at my avail to "connect the dots" for students to understand the concepts and skills within the teaching and learning.

Our education system is structured in a way that is segmented by the bell, the clock, and/or the subject. Although we have made attempts at integration of subject matter, it is still a challenge to do so in most of America's schools. We intentionally teach children how to read; although we may use a variety of strategies, the end result is to have students gain sight words, form sentences, and eventually read paragraphs and chapters in order to have the comprehensive skill of reading an entire book. The same is true, for the most part, with science. We often teach chemistry without the integration of biology or the life sciences. More often than not, we teach technology for technology's sake, as a separate elective class, and now with the onset of tablets or laptops for all students so that online learning can take place, we claim that technology is a met standard. Let me be very clear, for too long we equated technology with the simple use of a computer, when in fact the definition of technology is: the branch of knowledge dealing with engineering or applied sciences.

Let's take a deeper look at the word *applied*. There is a segmented understanding of theory and practice. However, think of a time when theory was absent of practice and practice absent of theory, such as when we are in academic settings where subjects were constructed for the sole purpose of theorizing and other courses were constructed for the purpose of practicing; that is, the practice of "applying" a skill, such as in mathematics and science. We use terms such as *experiment* and *replication*. Applied mathematics and applied science courses were often designed and provided for struggling students and historically marginalized students. This was an attempt to keep the abstract teaching and tracking of certain students in preparation for direct entrance into college or university post-secondary work. This is necessary to understand when describing and understanding STEM, as in some circles, this too can be misconstrued as a way to engage historically marginalized and/or students of color. The irony of this notion is that the education is the preparation for beyond schooling, and when we provide theory and practice simultaneously, learning becomes past, present, and future. Innovation motivates the leaner as well as the practitioner of the applied science, otherwise known as the workforce.

Currently in the United States, the economic statistics are a somber reminder as to the current state of education and the preparedness of our youth. There are 10.4 million jobs that are unfilled, 300,000 jobs are outsourced annually, and 36 percent of workers in the United States are part of the gig economy. Global outsourcing in 2019 was a $92 billion market. In addition, by 2025, IT outsourcing is expected to meet the $395 billion mark.[1]

How then do we begin to fill these unfilled professions with the citizenry of today and tomorrow? How do we communicate to parents, students, educators, and businesses the importance of discussions about robotic process automation, 5G plus, augmented reality, and cyber security? And we cannot have these discussions without the inclusion of the Mars mission(s). These technologies are here and growing.

However, the fact is that these discussions are longstanding, and we are seeing the results of well-intentioned advocates creating programs as best as they can with what they have. We should be intentional about the need of educating youth within a system that is steeped in STEM to provide a more congruent methodology that will prepare students beyond the classroom.

## Where Does STEM, Stem From?

STEM (Science, Technology, Engineering, and Mathematics) was first conceived and coined by Charles Vela, a Latino, somewhere between 1988–1991 at the Institute of Medicine of National Academy of Sciences and the Mitre Corporation in the Modernization of Federal Telecommunication Systems.

Vela affirmed six STEM principles:

1. The importance of deep and broad knowledge and an open mind.
2. The importance of an interdisciplinary approach and teamwork to tackle complex problems.
3. Most complex problems appear at the intersection (boundary and interface) of disciplines or components of a discipline and thus require an interdisciplinary approach.
4. The solution of complex problems is inspired by different bodies of knowledge, disciplines, and technologies.
5. The importance of differentiating between the essence of a problem (its complexity) and its appearance; its complexity from its complications.
6. The importance of knowledge appropriation as the

fundamental road to successful discovery and innovation.

Although there are internet sources that cite Judith Ramaly, an assistant director of education who came up with the SMET acronym and changed it to STEM between 2001 and 2004 while working for the National Science Foundation, I encourage you to do your own research regarding who coined and conceived the notion of STEM. I tend to believe Dr. Vela has the honor.

It is, however, important to note that there are a variety of attempts made to align STEM with that of the Morrill Act of 1862 to the response of Sputnik and the Nation at Risk report. It is safe to say there have been many renditions and attempts to address the need for education to focus on preparing students for careers of today and tomorrow.

The above information is significant in that in only 11 percent of the science and engineering jobs were held by "Black, Latinx and Native American workers."[2] Regardless of which side of the debate you stand as to whether education is meant to prepare students how to think and not what to think or about careers and not about specific careers, the argument stand moot when we continue to see the historical perspective and contrast between programs for learning and STEM as a systemic means to educate our youth.

## Policy, Practice, and Promise

Federal grants to implement policies and impact state and local education efforts have come in the shape of Perkins funds and supplemental funding opportunities. In addition, partnerships with the National Science Foundation and other non-profit and for-profit organizations have engaged in national, state, and local efforts. Why, then, are we still at the programmatic levels of sustainability with lasting results? Practice, practice, practice. We often misstep in the systems and structure for embedding STEM as a way of practice in education. Rather we move quickly to implement a program. The result can vary from state to state, district to district, and school to school and often is spawned and championed by an individual teacher.

Our current educational system is not structured for nimble pivotal adjustments, instead we create "programs." Programs to insert perhaps before or after a school day, which few students are able to experience. Or if there are pathways in which students are able to secure a specific course throughout the four years of high school, they too are often a marginalized few. Therefore, perhaps policies at the local level may have a greater impact if there was intentionality at the outset of a child's educational journey that provided the experience of a systematic approach for all rather than some. Allocation of priority spending may take on a whole new direction in educating our youth.

However, that is just one of the components of creating a systems view or approach to an integrated STEM education. The implementation factor is that of will and belief. There are plenty of ways to integrate Science, Technology, Engineering and Mathematics along with the Arts (STEAM), and they take time, planning, and most importantly, knowledge. As superintendents, we have the responsibility to provide our staff with the necessary tools to do the job they need to do. Simple, right? Not really. Many teachers get into teaching because they majored in the sciences, mathematics, literacy, or kinesiology. The point is, we must do a better job using every means necessary throughout our environment to reach students and connect them to world in which we all live. This is a shift in fundamental philosophy and, more importantly, in practice.

Priority spending goals perhaps could be a collaborative process between PreK–Higher Ed. This could possibly be a way to address staffing needs, staffing development for continued growth, and more importantly, a collective vision for what the 21$^{st}$-century student will need to impact us economically, environmentally, politically, medically, and technologically for the security and success of our nation and the globe. Many bright minds have the space and capacity to create networks and hubs of diverse thinking to solve current and future issues as long as we strive to construct better systems.

Practices of promise exist throughout our world, we are STEM and STEM advocates, and it behooves us to work with other leaders, decision makers, and partners outside of the education realm. The opportunity to craft a STEM education system is founded on data, results, and most importantly, a belief that all students are mathematics and science capable. As leaders of education systems, STEM is not a "nice to have," it is a "must have" for all students, and especially for our Black, Latinx, and Native American students who have been marginalized for too long. The practice of promise today is a promise kept for tomorrow!

# About the Author

## DR. MARIA ARMSTRONG

Dr. Maria Armstrong started a second career in education as a teacher of eleven years, counselor, AP, and principal and advanced through administrative positions as a director of English Language Learners, assistant superintendent of Curriculum and Instruction, a California superintendent, and an educational consultant for the Puerto Rico Department of Education Hurricane Maria Recovery efforts. She is now the Executive Director for the National Association of Latino Administrators and Superintendents in DC.

Dr. Armstrong champions equitable actions that provide students a brighter tomorrow. Based on her previous experience in the biotech industry, she is committed to ensuring that students are both college and career ready. Dr. Armstrong is a proud alumna of Azusa Pacific University, where she earned her master's in education with an emphasis on Counseling, and she is a recipient of the Influence award: *Honoring an alumnus whose investment of his or her profession and time has made a lasting influence on the character, development, or behavior of their students.*

She also holds a bachelor's in business and a doctorate in organizational leadership and received the Inspiration Award from the Sacramento Hispanic Chamber of Commerce with its Latina Estrella Awards.

Most recently, Dr. Armstrong received the Top 20 Female Leaders of the Education Industry from the American Consortium for Equity in Education.

Dr. Armstrong deeply believes that leadership and vision matter, but the life of an educated child matters more.

"What I love about kids is their pure honesty. They know when you're a champion for them or not. My goal is to provide hope, inspiration, and encouragement to genuinely care for and educate our children."

"What I am most proud of are my own children and grandchildren. My children saved my life, and education was my saving grace, for me and my family."

# THREE

# STEM–The GREAT Equalizer

Regina Armstrong

It is no secret we are living in a rapidly evolving world of technology. Yet are we really doing enough to prepare our students for their future success? Moreover, will our school systems keep up, especially when assessing students' STEM skills directly is not a requirement to graduate from high school? One way to begin to answer these questions is to ask whether your district has adopted a P–12 comprehensive STEM (Science, Technology, Engineering, and Mathematics) curriculum. The purpose of this chapter is to bring an awareness of how we must focus our efforts in updating our curriculum and measurement of student success based on what will be required of our students to compete and contribute to a global economy and society.

As a superintendent of schools, I have been challenged to implement a mission statement that creates the task of ensuring that students graduate from high school prepared for college and career ready. How can we possibly live up to the creed of our mission when STEM programs are treated as aspirational and not as a necessity? How can we implement STEM effectively when there is such a high demand for professional personnel in this area? How do you begin to address with our teaching staff that preparation for the 21st century is less about the information that can be found in a textbook and more about students' ability to critically think and make assessments focused on the world in which they live through STEM?

No surprise in my research, the United States, as great as we are as a country, continues to produce students who are lacking the necessary skills in the area of STEM. Kendall Hunt Publishing Company pointed out how studies have shown that math and science understanding will be a staple in all jobs moving forward (2016). However, US students are shown to be lagging in test scores on these two subjects, further proving the importance of STEM education for our youth and beyond. The question we must ask ourselves is, why are so many students in America's schools falling below the standards in these subjects? It is simple, as stated earlier, our course requirements fall short of students having to demonstrate proficiencies in these areas to graduate.

My awakening to how far my district was behind in preparing students for their future in this technological society was sparked by a visit to a school. During this visit, as I was observing classroom instruction, a teacher asked the class, "Who can tell me our nation's capital?" A student pulled out his cell phone and said, "Hey Siri! What is the capital of the United States?" In response the teacher became upset, had the student put his cell phone away, and reprimanded him. She reminded the student that the district had a "no cell phone use" policy. This was surely stated to impress me as the observer. She had no idea how disturbed I was that the student's rapid and accurate use of technology as a valuable resource was not embraced. This disturbance motivated me to redefine and reexamine how the district was truly preparing students to become global-minded citizens that could thrive in this era. It was an obvious conclusion that we had fallen short. Although we had made STEM an integral part of our after-school programs, we hadn't embedded this content into the day-to-day instruction. Therefore, only students participating in after-school activities were being exposed to STEM, equating to about 20 to 30 percent of the district's total enrollment. Consequently, this meant that more than 70 percent of the student body remained unexposed to STEM courses. We were leaving too many students behind.

That one-day observation awakened me to just how much the world had changed and how instructional delivery had not. The classrooms today still looked like they were following the same teaching methods that existed prior to the evolution of technology. It is time for change. Most educators agree that our current system of policies, procedures, and professional development are not designed to address and include the true meaning of STEM education. How can we expect teachers to begin to address STEM when we have not adequately trained them or provided required resources to effectively integrate STEM education in our schools? Now is the time to revamp our professional development plan and demand that colleges/universities do a better job of teaching, training and equipping future educators regarding the importance of STEM education. This could also be a unique opportunity, for most students are well ahead of their teachers in the use of technology. In case you have not thought about it, our students were born into a technological society. They are affectionately called Generation Z and are considered to be tech-centric. Generation Z is very comfortable in front of a computer, and most have never lived in a home without one, as well as cable, internet, and/or social media. They know nothing else besides being constantly connected to one another, albeit through phones, screens, and tablets (Cotrell 2020). So, why not capitalize on what comes natural for them in order to guarantee them a bright future?

It was that day, when I observed the student's response, that I made up my mind to lead the change I wanted to see and began to reimagine education for my students, such that our schools would increase opportunities for students to compete in a global society. In order to adequately prepare students to live a successful and productive life in the 21st century, STEM can no longer be elective or operational. Rather it must be an integral part required for all students from pre-kindergarten through grade 12. STEM education is the answer to transform our schools in our efforts to make our mission a reality and must be intentional in the way it is implemented. I was determined that our schools could no longer look at STEM as something you do in an after-school or Saturday program. STEM could no longer be something we do with our free time but must be something we live each day. Other schools and countries caught the STEM vision a long time ago; the time is now for my community to do so as well. It is the only way to give our students the tools they need to challenge and compete with their peers both near and far. To move forward with my district's transformation, I needed to be able to provide evidence to the stakeholders as to the need for the change.

Teachers need to be encouraged to allow students to use the basic technological tools to help them complete assignments.

Operating system assistants, such as Siri, should be friends not foes. Why are we still forcing students to solve long division instead of just handing them a calculator and teaching them how to use it? Does the new way of doing math really make sense, or is it a strategy being used to take away students' appreciation of math? Why is it no longer acceptable to teach students how to "borrow" from their neighbor? These questions illustrate reasoning as to why students are lacking in their ability to communicate effectively with their peers. For me, I am no longer focused on the why the district is behind, but I am taking proactive measures to create radical change. The implementation of a comprehensive P–12 STEM program will better guarantee our ability to increase the number of students reaching mastery and close the achievement gap.

As a child, I remember, two of my favorite cartoons were *The Flintstones* and *The Jetsons*. The irony of liking both of these shows is that they were the total opposite of each other. For those unfamiliar with these cartoons, both cartoons were made in the early 1960s; both dealt with family. However, while the Flintstones lived in a world that was a comical version of the Stone Age with machines powered by birds and dinosaurs, the Jetsons live in a comical version of a century in the future, with elaborate robotic contraptions and whimsical inventions. Growing up, neither show appeared to be realistic, and we viewed them simply for entertainment purposes. These shows illustrate the current state of STEM education.

The Flintstones' world was behind the times, and the Jetsons' concepts were unfathomable! But now, I realize my first exposure to a world of technology started many decades ago, as I tuned in to watch the wild imagination of George Jetson, his family, and his co-workers interact with their new gadgets inclusive of walking and talking robots. Who would have ever thought the Jetson would become our modern-day reality? Our schools must begin functioning with a Jetson-like mindset, but unfortunately, too many are still operating more like Fred and Barney (in the Flintstones). If you are not convinced there is such a stark contrast, just examine closely what is being taught in schools compared to the real world.

According to the National Science Foundation, students must develop their capabilities in STEM beyond what is currently happening in schools if they are going to succeed in this new information-based and highly technological society (National Science Foundation 2016). What we are doing now is simply not enough. STEM education develops students' critical thinking skills and provides hands-on learning opportunities for students to explore subjects in ways they may not have had the opportunity in the past (Kendall Hunts 2016).

The lack of STEM in schools is exacerbated in black and brown communities and students who live beyond the poverty line. We think that pushing diversity, equity, and inclusivity will help make students more productive citizen. This might be true; however, the only way we can bridge the gap for students of color and ensure they are more successful in life is by exposing them to STEM. Schools cannot carry the responsibility on their own. We must also educate our parents and guardians on the importance of exposing our students to higher level courses in math and science. This can be easily achieved by implementing district-wide STEM nights and other opportunities for the community to witness students' using their STEM skills in real-life situations. STEM education is not an option for us but a vital part of what we must do to secure our students' future success. Once we have increased students' exposures, we will be more effective in narrowing the achievement gap, providing them with the skills needed to compete in global society.

According to Kathryn Holmes et al. (2021), STEM fields are viewed as vital contributors to economic growth and innovation. Schools, especially those in urban settings, must begin to drastically change how they approached this area (Holmes et al. 2021). Why? It is simple. STEM skills and competencies are the key to increasing the number of people employable in these areas and are also viewed as highly transferable skills that increase employability in non-STEM sectors (Holmes et al. 2021). So, incorporating STEM in our classrooms will not only help our students become more employable, but it will also increase their critical thinking skills and overall academic achievement. STEM education will teach them how to adapt the concepts that they learn to various iterations of a problems or issues they will need to solve (Lynch 2019).

If we are truly going to use our schools to prepare students for success in life, we must start as early as possible pushing STEM. We can no longer deprive students the opportunity to use different types of technology (including cell phones) in the classroom. According to Holmes et al., engagement and interest in STEM subjects during primary and secondary schooling is pivotal to changing students' attitudes to pursue STEM subjects and careers (2021). In my humble opinion, educators stray away from STEM because of their lack of knowledge and fear. They fear our students will find out we have no clue what we are doing, given it is just not the world as we know it! No one prepared us for the advanced technological society we are living in.

Our schools can no longer afford to miss out on the most important aspects of teaching and learning for students' success—STEM. If we want to see students be successful in life and in their careers, it is past due for all schools to start mandating STEM education. I am not sure why it is taking so long for us to do so. The pandemic has shed more light on the readiness gap. Computers are now a necessity, not a luxury! The pandemic, as horrific as it is, should also serve as a wake-up call to let schools know that we must begin to change the way we educate students, NOW! If we do not integrate STEM as a part of our lifelong learning and totally immerse STEM into our curriculum, professional development, board policies, practices, and missions, history will show that our schools were negligent because they breached their duty to develop students into productive members of society. Let us not only make students smart using measurements from yester-day's standards but let us also make certain students are resourceful and prepared for today and their future!

## References

Cotrell, Sarah. 2020. "A Year-by-Year Guide to the Different Generations and Their Personalities." *Parents*. https://www.-parents.com/parenting/better-parenting/style/generation-names-and-years-a-cheat-sheet-for-parents/.

Holmes, Kathryn, Erin Mackenzie, Nathan Beger, and Michelle Walker. 2021. "Linking K–12 STEM Pedagogy to Local Contexts: A Scoping Review of Benefits and Limita-tions." Centre for Educational Research, Western Sydney University, Penrith, NSW, Australia. https://www.fron-tiersin.org/articles/10.3389/feduc.2021.693808/full.

Kendall Hunt Publishing Company. 2016. "The Importance of STEM Education." https://k12.kendallhunt.com/blog/im-portance-stem-education.

Lynch, Matthew. 2019. "Seven Benefits of STEM Education." https://www.theedadvocate.org/7-benefits-of-stem-education/.

National Science Foundation, 2016. *Science and Engineering Indicators* 2016. https://www.nsf.gov/statistics/2016/ns-b20161/uploads/1/12/chapter-2.pdf.

# About the Author

## REGINA ARMSTRONG

Regina Armstrong is the Superintendent of Schools in the Hempstead Union Free School District. She obtained her bachelor's degree in Elementary Education from Queens College, Flushing, New York, and attended Hofstra University to pursue a Master of Science in Reading. The information she learned became invaluable in developing her teaching practices that led her to receive many awards for her students achieving their yearly targets in Reading and Math.

Believing in being a life-long learner, Regina obtained her Professional Diploma from Queens College in August 1998 and is currently pursuing her Doctorate at American College of Education, Indianapolis, Indiana in Education Leadership with a focus on Curriculum and Instruction. In June of 2017, her life's work was recognized by the Hempstead Branch of the NAACP, were she received the Education Award. Regina counts it a privilege to have been employed in the Hempstead Union Free School District (HUFSD) for the past thirty-one years, serving in the following positions: teacher, curriculum specialist, assistant principal, and principal.

For three years, she served as the Interim Superintendent of Schools for HUFSD, and on June 17, 2021, she was appointed Superintendent of Schools. Although it is the most challenging accomplishment of her career, it is also the most rewarding. She considers it an honor and is humbled by the opportunity to lead the District in this capacity.

As much as Regina loves and appreciates her career as an educator, she knows none of her accomplishments would be possible without the grace of God. She has been a member of the Emanuel Baptist Church in Elmont since 1978. For the past eleven years she has served as the Supervisor of the Emanuel Youth's Church and an adult Sunday school teacher. Her church activities also extend to her local district association, the Eastern Baptist Association of New York, Inc. She was appointed the General Supervisor for the Maggie B. Greene Youth and Young Adult Ministry and serves as a member of the Scholarship Commission and Special Events Team. She feels blessed that God has given her a servant's heart and uses her gifts to teach, witness, lead, and support Kingdom building.

Regina feels blessed and fortunate to have a strong support team made up of her loving mother, Corine, her father J.R. (from Heaven), four sisters—Phyllis, Patricia, LaVern, and Michelle—eleven nieces and nephews, and several loyal friends and colleagues. Her two favorite scriptures are: *"No weapon formed against me shall prosper* (Isaiah 54:17a) and *"God can do all things exceedingly, abundantly, above all we can ask or think according to the power that works in us"* (Ephesians 3:20). Her life is a testament of both.

# FOUR

# Educational Equity and STEM

## THE LACK THEREOF

### Dr. Fadhilika Atiba-Weza

Approximately twenty-five years ago, an education task force was created to visit and evaluate low-performing schools in New York. The team members were charged to, among other things, examine the following: How were the schools performing? What factors contributed to the condition of the schools? How was the school community addressing the issue of low performance? What resources were needed to advance the improvement process? What were the prospects of the school sufficiently improving to meet the State's standards of acceptability?

The task force visited a regional vocational high school, now called a career and technical education (CTE) school. The team noticed that the black and brown students were overwhelmingly placed in "non-technical" classes. Cosmetology, hair dressing, nail salon, and secretarial science were offered to the girls. The male students, on the other hand, were steered toward carpentry, landscaping, and other programs that were deemed low-status programs. The other students—Asian and white—were primarily placed in the "more technical and academic courses"; that is, avionics technology, automotive technology, computer programming, medical assistance, medical technology, and similar courses of study.

A meeting with the school's leadership staff resulted in them reporting that they were not aware of the correlation between race and ethnicity, on the one hand, and student placement and enrollment, on the other. When shown the data and asked to explain the disparate approach to class assignments, the counselors were flummoxed. Interestingly, the task force members arrived at the conclusion of the relationship between ethnicity and course assignments by looking at the classrooms. No data was necessary. In other words, the enrollment pattern and class assignments were visible to everyone who visited the school.

Since that time, there has been no significant improvement regarding the number of black and brown students in Science Technology, Engineering, and Mathematics (STEM) classes and careers. Landivar (2013) noted that Blacks and Hispanics have been consistently underrepresented in STEM employment. In 2011, 11 percent of the workforce was Black, while 6 percent of STEM workers were Black (up from 2 percent in 1970). She also found that Hispanics comprised 7 percent of the STEM workforce. Several years later, Martinez and Gayfield (2019) found that conditions had not significantly improved. They reported that Black men and women comprised 4.1 percent and 2.2 percent of the STEM workforce respectively, while Latino men and women, comprised 5.3 and 1.7 percent of the STEM workforce.

According to the Pew Research Center (2018), Blacks and Hispanics are underrepresented in science, technology, engineering, and math jobs, relative to their presence in the overall US workforce, particularly among workers with a bachelor's degree or higher.

Sixty-seven years since the historic case of *Brown vs the Board of Education of Topeka Kansas*, the quality of the education for black and brown students lags that of the mainstream, STEM being among the areas of greatest challenge. Many guidance and career counselors, like those who worked at the aforementioned CTE, apparently fail to grasp the economic and social value of STEM for all, and more so the role that African Americans have played in the scientific field.

Educational leaders are therefore tasked with, among things, the challenge of making sure that teachers, school counselors, and other professionals appreciate the role of black and brown people in the development of science and technology. In addition, they must insist that all children are afforded the opportunity of receiving a first-class education, irrespective of the subject and the student to whom it is taught. Too often, the implicit bias of educators blinds them to the reality of the damage which they exact on students. Staats (2015) found that many well-minded teachers are unaware of the consequences of implicit bias. Adults must be intentional and proactive in addressing and/or avoiding the negative consequences of implicit bias. The great African American human rights activist and leader Malcolm X recounted the incident in which one of his teachers told him that he was better suited for carpentry than he was for law. This occurred even though the young Malcolm was a bright and articulate student. Somehow, the teacher seemed to have been unable to see him as an attorney.

Policy makers and others need to champion the work of the great African American inventors and scientists so that, among other things, our young people can see images of role models who look like them.

A casual reading of history reveals a rich list of Blacks in the STEM field (Fouché 2003; Gubert et al. 2002, among others). African Americans are not new to or incapable of handling the rigors that are associated with the pursuit of careers in the STEM fields. It is not the purpose of this chapter to ascribe motives to people's actions. The challenge is for us to confront the situations as they occur and develop strategies for addressing the shortcomings.

This is critically important during this period in our country's history, a time during which many are marginalized and attempts to address such marginalization and its impact are met with hostility.

School leaders must work with all staff, particularly those like the counselors who "steered" the black and brown students away from the more technical and academic STEM courses. Their actions are demonstrative of the damage that is inflicted on black and brown students by well-minded but misguided educators.

*Why the fuss about STEM?*

Should we not look at all subjects and courses of study along with the resultant careers as important? All subjects, courses, and careers are important. However, the rapidly evolving social and economic landscape requires us to engage our students in ways that enable them to be successful in the 21st century. Several organizations have reported that today's economy and workforce require competency and mastery of

STEM in order to move our society forward, and moreover for it to be competitive in the 21st century.

The Joint Economic Committee of the United States Congress (2021) argues that "increasing participation in STEM fields among women and marginalized communities is one way to drive future innovation. Research has shown that if more women and Black Americans were engaged in the technical innovation that leads to patents, U.S. GDP per capita could be 0.6 to 4.4 percent higher."

The Committee also argues that a barrier to such involvement is the digital divide that has plagued this country for much too long:

> This digital divide is further exacerbated by racial disparities, with 25 percent of Black teens reporting that they are sometimes or often unable to complete homework assignments because of a lack of a reliable computer or internet connection, compared to 13 percent of White teens and 17 percent of Hispanic teens.73 As schools have relied on online learning during the pandemic, children's ability to fully participate in their learning and grow their human capital has differed based on their family's economic and racial demographics.

Everything that we do seems to have a STEM connection, whether it is driving your vehicle, writing a letter, or taking a shower. Not only are these anecdotal statements evidential of the ubiquitous nature or STEM, the US Department of Commerce (2017) states that:

- Employment in STEM occupations grew much faster than employment in non-STEM occupations over the last decade (24.4 percent versus 4.0 percent, respectively), and STEM occupations are projected to grow by 8.9 percent from 2014 to 2024, compared to 6.4 percent growth for non-STEM occupations.
- STEM workers command higher wages, earning 29 percent more than their non-STEM counterparts in 2015. This pay premium has increased since our previous report, which found a STEM wage advantage of 26 percent in 2010.
- Nearly three-quarters of STEM workers have at least a college degree, compared to just over one-third of non-STEM workers.

Given the importance of STEM, it goes to reason that all sectors of our population ought to be involved in its development, should this country want to be economically competitive on the world's stage.

## References

Fouché, R. (2003). *Black Inventors in the Age of Segregation: Granville T. Woods, Lewis H. Latimer, and Shelby J. Davidson.* Baltimore: The Johns Hopkins University Press.

Fund, C. and Parker, K. (2018). Blacks in STEM Jobs are Especially Concerned about Diversity and Discrimination in the Workplace. Report of the Pew Research Center.

Gubert, B., Kaplan, M.S., and Fannin, C.M. (2002). *Distinguished African Americans in Aviation and Space Science.* Westport, CT, Oryx Press.

Landivar, L.C. (2013). *Disparities in STEM Employment by Sex, Race, and Hispanic Origin. American Community Survey Reports.* September.

Martínez, A. and Gayfield, A. (2019). The Intersectionality of Sex, Race, and Hispanic Origin in the STEM Workforce. Social, Economic, and Housing Statistics Division U.S. Census Bureau SEHSD Working Paper Number 2018-27.

Staats, C. (2015). Understanding Implicit Bias. What Educators Should Know. United Federation of Teachers.

United States Department of Commerce. (2017). STEM Jobs: 2017 Update.

United States House of Representatives. (2021). Joint Economic Committee Congress of the United States on the 2021 Economic Report of The President.

X, M. (1965). *The Autobiography of Malcolm X.* New York: Ballantine Books.

# About the Author

## DR. FADHILIKA ATIBA-WEZA

Dr. Fadhilika Atiba-Weza is the Executive Director of the National Alliance of Black School Educators (NABSE). He is an educator with more than forty years of experience, having previously worked as a classroom teacher, building administrator, district administrator, and university faculty member.

At NABSE he provides administrative leadership as the Chief Executive Officer and National Spokesperson of the organization.

A former school superintendent, Dr. Atiba-Weza has been involved in many educational reform and school improvement projects. Additionally, he has served on the boards of the National Alliance of Black School Educators (NABSE), the New York State Council of School Superintendents (NYSCOSS), the New York State Association of Supervision and Curriculum Development (ASCD), among others.

Dr. Atiba-Weza is a social activist with involvement in many African, African American, and Caribbean organizations.

He received his Bachelor of Arts Degree in Psychology from the City College of the City University of New York, a Master's Degree in Education Guidance and Counseling from the Brooklyn College of the City University of New York, a

Master's Degree in Educational Administration and Supervision from the City College of the City University of New York, a Master of Education Degree in Organizational Leadership from Columbia University's Teachers College and his Doctor of Education Degree from The Sage College of Albany.

He is a member of many professional and civic organizations including life membership in the National Alliance of Black School Educators and the National Association for the Advancement of Colored People, along with membership in the American Association of School Administrators, the American Educational Research Association, the Caribbean Studies Association, and the Sigma Pi Phi professional fraternity.

An avid gardener, jazz enthusiast, and golfer, Dr. Atiba-Weza resides in the New York Capital District.

# FIVE

# Coding

## THE NEW UNIVERSAL LANGUAGE

Mark D. Benigni, Ed.D.

Many STEM fields include computer programming and interaction with devices as well as the need for emotional intelligence and collaboration with humans. Technology will continue to play a more significant role in the global economy, and our ability to communicate and interact with machines will be at the forefront. Just as we recognize the importance of collaboration and compassion, we must also teach our students how to communicate with devices through the language of coding.

We need to build a culture where today's graduates leave our high schools prepared for the challenges of the real world. We must ensure that we provide opportunities for our students to become learners, thinkers, advocates, and collaborators. Can students direct their learning, persevere, and take risks to find happiness and achievement? Can students think critically and problem solve? Do our students embrace diversity and show empathy and respect for others? Are our students able to achieve short-term and long-term success? Can our students communicate effectively and resolve conflicts with their colleagues...and with their robotic counterparts?

## Coding for Academic Success

"Good morning, Alexa. What will the weather be like today?" That may be a typical first question of the morning for many of us. Next up is grabbing the cell phone to start the coffee brewing. As one sips the warm coffee, they may be programming their washing machine to start a load of laundry at 3:00 p.m. so that it is ready for the dryer after the workday. With another cup of coffee, they may realize that the coffee creamer is down to the last drop. No worries—just scan and add it to the refrigerator-generated list. Then use voice activation on the vehicle's daily companion autopilot to begin your journey to the office. You feel safe as you depart your home for the day because your home security doorbell provides a consistent extra set of eyes.

Whether you live this way or just utilize your smart TV for Netflix, Prime Video, and Hulu, you can relate. Others may just be too busy being guided by their Apple watch and reminders to move. Either way, no one can deny that science, technology, engineering, and math will continue to change the way we live, how we think, and the jobs of the future.

Where will the future take us? What will be the jobs of tomorrow? According to Code.org, 71 percent of all new jobs in STEM are in computing, and only 8 percent of STEM graduates are in computer science. We also know that our school systems continue to struggle to fill math and science teacher openings. We must be creative in our strategies to ensure and insist that STEM plays a part in our students' learning.

STEM curriculum can no longer begin in our high schools but must start with our youngest learners in kindergarten as we build a culture for the future. The process of authentic creation allows all students to make things that matter to them and in the medium that interest them most. These personally relevant projects are laying the foundation of computer science and personalized learning. As students progress, they will become versed on the internet, navigate digital information, learn how to program, and build applications. All of this will be done with data privacy, security, and safety in mind.

Educators and community leaders are wrestling with redefining the jobs of tomorrow. As former manufacturing factories sit vacant and become targets for crime and undesirable activities, our business leaders are also searching for answers. By embracing Science, Technology, Engineering, and Math, together we can create the STEM Century. That is how we will redefine our communities and create the jobs of tomorrow. To bring these changes to fruition, we must start by making the necessary educational changes that our students have requested, both in our schools and with our virtual partners.

## Creating a Culture of Innovation

In the Meriden Public Schools, we have been recognized for our digital transformation and culture of innovation. While STEM education has always been a focus area, we now realize that our work must start with our youngest learners if we are going to lay the groundwork for this cultural change. That is why we have partnered with Sacred Heart University and Code.org through Project {FUTURE} to bring computer science to our elementary schools. Through a U.S. Department of Education Innovation and Research Grant, we launched a comprehensive coding program for all our elementary students starting in kindergarten. We are well on our way with a high-quality Code.org curriculum in place and instructional coaching for all of our teachers.

Coding teaches the skills of problem-solving, persistence, creativity, collaboration, and communication. These are the skills our students need for academic success in reading, mathematics, science, social studies, and life. Coding is not another activity jammed into a busy schedule. It allows students to develop necessary skills and dispositions through computer science. According to @teachcode, students will learn to "value and expect mistakes as a natural and productive part of problem-solving, experiment with new ideas and consider multiple possible approaches and mediate disagreements and help teammates agree on a common solution."

Additional STEM and coding experiences are offered through the Meriden Public Schools' partnership with The Bushnell Center for the Performing Arts Digital Learning and Community initiatives. Through a grant with MusiQuest, The Bushnell has provided our students an opportunity to enter the world of music and embark on a creative new way to learn coding through music. The focus of MusiQuest is to allow students to create their own music by building a song from scratch. The lessons are interactive, hands-on, and fun for students. They can play with over seventy-five instruments as they create their own musical piece. MusiQuest's curriculum integrates music fundamentals with math, science, nature, literacy, social-emotional learning, and more.

As we work to ensure that all of our students graduate from high schools with a college and career plan in place, we know that data entry positions have been great career starters for our students. With coding universally incorporated in our younger grades, it was time to focus on providing creative opportunities for our high school students. We looked no further than our terrific business and community partner, The Bushnell Center for the Performing Arts.

As part of The Bushnell Digital Learning Workforce Development Program, The Bushnell hosts an eight-week summer internship on digital design and production. Aspiring high school students who wanted to learn more about coding, art, and writing were encouraged to apply to The Bushnell program. Students in the program create their own video games. The intensive lessons include Adobe Creative Cloud (Photoshop), Illustration Writing in HTML (using the construct game three engine), and Collaboration Across the Web (writing narrative documents and preparing budgeting and marketing plans).

As we create schools and a culture that prepare students for today and tomorrow, coding must be embedded into all content areas. Coding should be part of what we do, not just another add-on.

## Ten Thousand Lines of Code in One Night

It was a chilly, windy, and drab early December evening when I headed to one of our elementary schools for our district's Hour of Code Event. After a long day at the office, I thought I would pull right up to the school and make a brief appearance at an event that would probably have limited attendance and even less energy and excitement. I knew we had promoted the event with flashy posters and personalized messages to families.

When I arrived at the school, I was shocked to see so many cars. I struggled to find a parking space. What else was going on at the school? Youth basketball? The community choir? A neighborhood association meeting?

*Figure 1. Hour of Code Poster, Meriden Public Schools*

After finding a parking space between two large SUVs, I entered the school with limited expectations and plans for a quick stop and a brief staff thank you. What I saw amazed me! The gym was filled with student-made racetracks, Ozobot Bits, and Ozobot Evos circling the large student-developed race-tracks. As parents and family members cheered on like they were at the Daytona 500, our students' excitement, engage-ment, and energy picked me up after a tough, long day. Then I visited the Media Center to see the next Hour of Code Station, and there I saw parents and younger students working together to program the Code & Go Robot Mouse.

As I saw one student holding the mouse and scanning the Media Center, I knew it was now my turn to have a little fun. So I approached the student, who gave me thorough program-ming directions and watched as I tapped the forward arrow five times before pushing the right arrow three times. I anxiously awaited for my mouse to move. Maybe it needed new batteries? As it sat idle, the student came over and gently touched the middle green button for me, and just like that, my mouse began the journey I had programmed for it.

At another station, students selected one of Code.org's tutori-als. They used block coding to code their own Dance Party, design Flappy Code games, or program characters, such as *Frozen*'s Elsa and Anna or *Star Wars*' Leia and Rey, to complete a series of tasks. By the end of the evening, our students had written over ten thousand lines of code. Students and parents were also introduced to the wonder of virtual reality as they explored undersea adventures and ancient wonders of the world through VR headsets. As students coasted through their activity stations, teachers signed off on their Hour of Code Passports.

After sharing some laughs and enjoying some hands-on activities, I exited to the cafeteria only to see that Botley® was a big hit there! Botley was a larger robot that had the ability to push and move things. What indeed had moved me from this night was my thinking about how the coding of robots could play a role in our district's STEM efforts and promote a culture of learning.

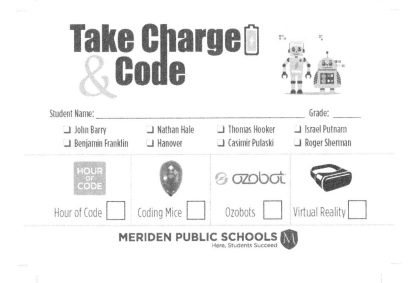

*Figure 2. Hour of Code Passport, Meriden Public Schools*

## Jump-Starting Transformation

In Connecticut, Governor Ned Lamont recently announced a collaboration between the Connecticut State Colleges and Universities, the Connecticut Office of Workforce Strategy, and Amazon WebServices to train more than two thousand residents for careers in cloud computing. All Connecticut state and community colleges will be offering courses and certificates that align with the skills needed to gain employment in high-demand technical careers. In addition, colleges will work with interested high schools to ensure these students have access to cloud computing curricula and resources. Governor Lamont stated, "We know that a workforce rich in tech talent is one of the keys to Connecticut's future economic success."

Our graduates will need to embrace imagination, innovation, and technology integration at an early age to be ready to compete in the STEM Century successfully. We need to raise 21st-century graduates who design organizations that imagine what is possible, support innovation, and inspire employees. Together, we can jump-start transformation by ensuring that all of our students learn the universal language of coding. Only then will society truly recognize and maximize the benefits of what technology can do to support learning and work.

# About the Author

## MARK D. BENIGNI, ED.D.

Dr. Mark Benigni has been an educator for over twenty-five years and is in his twelfth year as superintendent of the Meriden Public Schools. Dr. Benigni served as a teacher, principal, four-term mayor, and has authored over thirty-five articles in educational journals. His book, *Mentoring Matters,* was published by Rowman and Littlefield Education (2011). Since Dr. Benigni's arrival in 2010, the Meriden Public Schools has increased student academic achievement and growth and created schools where students and staff want to be.

Dr. Benigni has presented at national venues, taught doctorate classes, and currently serves as president of the Connecticut Association of Public School Superintendents, after having spent many years as co-chair of the Connecticut Association of Urban Superintendents.

Dr. Benigni was recognized with AASA's 2020 Digital Superintendent of the Year Award, a 2015 Education Week Leader To Learn From Designation, a Social Justice Award from the CT Coalition to End Homelessness, the William A. Yandow Educator Award from the Bushnell, and he was a recipient of the Ten Outstanding Young Americans Award by the United States Junior Chamber of Commerce. Under Dr. Benigni's leadership, the Meriden Public Schools has received awards from ISTE, CoSN, The Learning Counsel, NSBA, Digital Promise, District Administration, Edutopia, HundrED, Google, and the International Center for Leadership in Education.

Dr. Benigni grew up in Meriden, and his children Bria and Blake attend the Meriden Public schools. He welcomes the rich diversity of an urban community and is committed to providing exceptional learning opportunities for all students in the Meriden Public Schools.

# SIX

# A STEM Education for All

## PROVIDING THE VILLAGE A STEM EDUCATION TO SUPPORT THE 21ST-CENTURY LEARNER

### Racquel Berry-Benjamin

My first professional connection to STEM was early in my education career as an elementary teacher when I was reminded of its importance. Luckily for me, and my students, teaching STEM came naturally for me, as the concepts are aligned with my philosophy on how subjects should be taught (interconnected) and how students learn best (through active engagement).

Students should *experience* learning, and a STEM education offers countless experiential learning opportunities through experiments, inquiry- and discovery-based methods, projects, and solving real-world problems that make learning interesting, relevant, and fun.

As an elementary educator, I was charged with teaching all academic disciplines. Using a project or problem as the central base for teaching made it easy to connect all disciplines and teach them through the project/problem. Reading, writing, and discussing the project/problem made sense to students because classroom lessons were interconnected to the same central point.

While serving as a mathematics coach, I developed and ran the school's first-ever Math Bee Competition for students to demonstrate their advanced numeracy skills. Students felt a sense of community being a part of a group of like-minded peers and a sense of achievement in applying their math skills during the rigorous competition.

As I progressed through my career, my connection to STEM shifted as my professional role changed from schoolteacher to district administrator, and later to deputy superintendent. On the district level, my connection to STEM was in an oversight and support role, ensuring resources and supports were in place to assist teachers, coordinators, and directors operating STEM programs.

In my current role as Commissioner of Education (Chief State School Officer) for the Virgin Islands Department of Education, my connection to STEM has once again evolved. I now have the responsibility of creating the vision and leading the transformation of a public education system so that 21st-century learners are prepared to successfully live, work, and thrive in a 21st-century world. STEM remains at the core of my vision to transform the Virgin Islands public education system, as it relates to what students are taught and how they are engaged in educational experiences.

## My Modern Approach to Education Transformation

One afternoon, my nephew, three years old at the time, was in my family room playing with his Lego blocks. In passing, an adult family member said to him in local, colloquial dialect, "Zyan, I see you is a Lego Man."

My nephew, very articulately and without hesitation, responded, "No, I am an engineer."

Everyone in the home erupted in tear-inducing laughter. Although my nephew is known to often surprise us with his extensive vocabulary, this interaction took the cake and astonished us all.

One of the great philosophers whose work greatly inspires me, John Dewey, once said, "If you teach today's children the way we taught yesterday's, we rob them of tomorrow." I echo these sentiments and wish to add, *If we speak to today's children the way we spoke to yesterday's, we rob them of tomorrow.* With this in mind, I have identified some of the primary sources that have influenced my approach to transforming schools, communities, and ultimately, the educational outcomes of students:

- Professional experience teaching and leading in education
- Research on the positive effects of experiential learning
- Research on the positive effects of project/problem-based teaching
- Research that indicates a STEM education is essential for 21st-century students to be successful in work in the 21st century
- Personal, real-world experiences

According to the US Department of Education (2015),

Every child is imbued with a sense of curiosity and wonder. They are born scientists, engineers, and creators ready to discover the world at every turn. The goal of education should be to sustain this engagement throughout a lifetime. We need our partners in both the public and private sectors to take up the challenge and help us to answer these and other questions: What types of active learning strategies work best for young children? How do we encourage curiosity and creativity that lasts a lifetime? What roles do computer science and computational thinking play for young children? How do we best leverage technology in the classroom? Most importantly, how do we prepare educators?

## Future Work Skills

*McKinsey February 2021 Data: Manual Labor 4 percent used in the future*

COVID-19 has made it glaringly evident that technology, artificial intelligence, and remote work will increase drastically by the time today's children make up the workforce. According to McKinsey Global Institute (2021), workers will need to learn more technological skills in order to move into occupations in higher wage brackets. The research suggests that workers that earn the highest wages will spend 26 percent of time applying technical skills in comparison to 4 percent of their time spent applying physical and manual skills and 6 percent of their time applying basic cognitive skills (see Figure 1).

## The Future of Work Is Dependent on Workers Having a STEM Education

Now is the time to create STEM communities and provide every student a STEM education. The pandemic has afforded the opportunity to reset, redesign, and reengineer education systems. The greatest challenge is predicted to be uprooting cultural norms and mindsets and enduring the process of change. Without a doubt, change is uncomfortable. It naturally induces feelings of anxiety and stress, as it may require new attitudes and mindsets toward teaching practices, modes of instruction, student engagement, and learning as a whole.

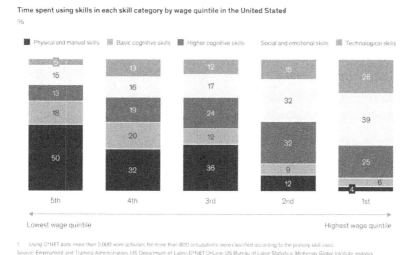

Time spent using skills in each skill category by wage quintile in the United States
%

■ Physical and manual skills ■ Basic cognitive skills ■ Higher cognitive skills Social and emotional skills ■ Technological skills

*Figure 1. The Future of Work After COVID-19 (McKinsey Global Institute, 2021)*

Embarking on a system-wide implementation of STEM education, where it serves as the first mode of instruction for students, is no simple task and requires years of consistent knowledge-sharing until full implementation. There is countless supporting research that suggests:

- a STEM education is essential for the 21st-century learners of today to be successful in jobs of the future; and
- parental engagement and community support are major factors that contribute to student success.

Students should experience STEM everywhere—in school, at home, and in their communities. Providing "all students" access to STEM increases the likelihood of all students receiving a STEM education. Providing access to "parents and the community" increases the chances of students receiving a sound STEM education and having a well-rounded STEM experience.

The US Department of Education, in its article entitled, "Communities Come Together to Support STEM Education," stated that "cultivating a creative workforce that is ready to step into science, technology, engineering and math (STEM)-related fields is vital. Students need technical knowledge in these subjects, as well as critical-thinking, problem-solving, and analytical skills; these tools will prepare them for tomorrow's jobs....A STEM learning 'ecosystem' creates connected learning opportunities for students throughout their community, both within and outside of school" (US Department of Education, 2015; see Figures 2 and 3).

Parents and community members must speak the language of STEM, but to do so, they must first know the language. Providing access to STEM for adults around children ensures they have the education and resources needed to be able to assist students. It opens the door for teachers, parents, and community members to engage students in STEM-based lessons and activities.

Providing a social studies teacher access to STEM training and resources can open a level of insight that teacher had not previously considered. Educating parents and community members on STEM will build their confidence and serve to eliminate fear around STEM. They will feel empowered and like partners in the education of their children, which is a goal we strive for in education.

Providing access is the key to breaking barriers and shifting mindsets around STEM education. All students can and are required to learn through a STEM-based approach to succeed in jobs of the future that do not yet exist.

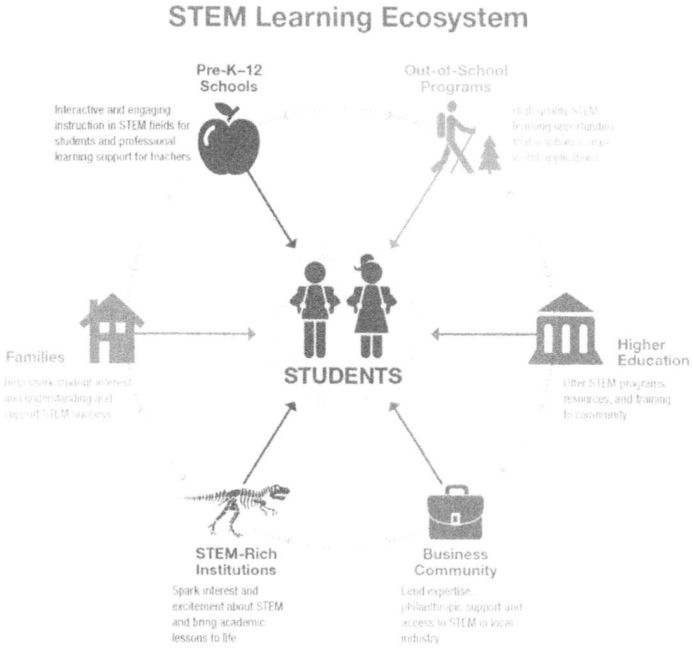

## STEM Learning Ecosystem

Figure 2. Diagram illustrating the potential anchor organizations of a STEM learning ecosystem (US Department of Education, 2015)

The information in this chapter is crucial to educators, students, and communities because it requires the commitment and collaboration of everyone to transform education through STEM. All stakeholders must understand STEM is the essential ingredient in a 21st-century learner's educational experience, and it must be treated with importance.

*Figure 3. Learning Ecosystem Diagram inspired by and adopted from the US Department of Education (USDE) (2015) and enhanced and adapted by Racquel Berry-Benjamin (2021) to support the content of this chapter. The USDE created the diagram to highlight the potential anchor organizations of a STEM learning ecosystem.*

## Consistent STEM Implementation Is Key

The solution to the challenge of not enough children having a STEM education is to implement STEM throughout the entire education system. Develop a strategic plan for a systemic integration, and remain consistent in implementation.

Regardless of political influence, change in administration or education leadership, the plan must be executed with fidelity to the end.

Involve the community so that they understand the importance of a STEM education and understand their role in the big picture to educate 21-century learners. Provide them access to information, programs, and STEM services so that they have the resources and feel supported in the learning process. Technology is a powerful tool to help educate the masses. Online platforms such as Zoom and Microsoft Teams allow multiple persons to participate at once. Recorded information can be replayed as many times as needed to gain an understanding of the information. The use of technology helps to ease the learning process and supports the various schedules of adults in the community; they log on at a time that is convenient for them.

## Committing to Increasing STEM Access

It is evident, that my approach to STEM has evolved. It began with me as a practitioner in the classroom, then an overseer and support to other practitioners, and now as a system-wide implementer and advocate.

The moment that expanded my understanding of the importance of STEM in my community occurred some years ago, as a result of several experiences. Given the aged infrastructure of our beautiful islands, we experience seemingly insurmountable safety and quality-of-life problems, such as regular power outages and deficiencies in the foundational integrity of roads and buildings, to name a few. Many of our infrastructural challenges have required engineers and/or a variety of other skilled professionals, who, at the time, would fly in from neighboring

Puerto Rico and/or Florida. This resulted in residents having to wait with uncertainty and anxiety.

We have made tremendous progress in recent years by increasing the number of skilled workers in the Territory. This has largely occurred as more professionals relocate back home or new people take up residence. As Commissioner of Education, I am committed to systemically implementing STEM using a system-wide approach to provide access to all students, persons, and organizations supporting the education of children. The K–12 role in this movement is to inspire students to explore careers in STEM by providing them the opportunity to learn through STEM-based projects and problems. A goal is also to increase post-secondary opportunities in STEM for students.

## The Only Way to Go Is Up

We are twenty-one years into the 21st century, and a STEM education is essential for 21st-century learners. I have seen where STEM has created some of the most perseverant problem solvers and has proven to be essential to the success of students. The COVID-19 pandemic has, in many ways, leveled the playing field and identified the ways in which our current education systems are not serving our students. The only way to go is up. Let's build new systems of education with the necessary content, instructional methods, and educational experiences that are pertinent for students to be successful in post-secondary education, in the workforce, and in life.

This is an opportunity that may never come around again in a lifetime. I encourage all educational leaders, legislators, education board members, labor leaders, teachers, parents, students, and members of communities to be courageous and support a system-wide infusion of STEM in schools across all grade levels and to commit to doing your part.

I leave you with one of my favorite quotes from the great Abraham Lincoln, "The best way to predict the future is to create it." Join the movement to transform today's learners into tomorrow's leaders, our 21st-century learners are depending on you.

## References

McKinsey Global Institute (2021). The Future of Work After COVID-19. https://www.mckinsey.com/featured-insights/future-of-work/the-future-of-work-after-covid-19

US Department of Education Office of Elementary and Secondary Education Innovation. (2015). Communities Come Together to Support STEM Education. https://oese.ed.gov/2015/11/communities-come-together-to-support-stem-education/

US Department of Education Office of Elementary and Secondary Education Innovation. (2015). STEM Education: A Case for Early Learning. https://oese.ed.gov/2015/12/stem-education-a-case-for-early-learning/

# About the Author

## RACQUEL BERRY-BENJAMIN

Hailing from the beautiful United States Virgin Islands, Racquel Berry-Benjamin serves as the Commissioner of Education (Chief State School Officer) at the Virgin Islands Department of Education. As the youngest education professional to serve as Commissioner in the U.S. Virgin Islands (USVI), Cmmr. Berry-Benjamin has a bold and future-facing vision of transformation for public education in the U.S. Virgin Islands.

With a well-rounded twenty-year career in education, Cmmr. Berry-Benjamin has a thorough understanding of how students learn best, what quality teaching and effective school leadership looks like, and the functions of educational systems after serving as a paraprofessional, teacher, mathematics coach, district administrator, state director, part-time professor, deputy superintendent, and Commissioner of Education.

Coupled with her professional experience is her personal experience of matriculating through the Virgin Islands Public Education System and her experience as a mother of children who also went through the public education system in the Virgin Islands.

In Cmmr. Berry-Benjamin's chapter, you can expect to learn about providing students, education professionals, parents, and community members access to a STEM education to transform educational systems and ultimately achieve the desired educational outcomes for our students. Her hope is that as a result of this work, educational systems and communities transform to provide everyone in the community access to STEM so that students get a well-round 21st-century education to thrive in the 21st-century world.

# SEVEN

# The Convergence of PBL and STEM

## EDUCATIONAL DEVICES TO POSITIVELY IMPACT SOCIETAL ILLS

Melvin J. Brown, Ed.D.

Both project-based learning and STEM (science, technology, engineering, and math) education have been tenets of our school district—in some way, shape, or form—for over twelve years. We have partnered with agencies to provide training and support and to work with us to develop the first officially designated K–12 STEM pipeline in the state of Ohio. It has given us context and perspective as we have navigated the growing disparities that have existed in education for decades. While implementing STEM and project-based learning (PBL) in some of our schools, we also discerned that, by only doing so in select schools, we were contributing to the divide among students with means and those without. Our "STEM schools" have always been schools of choice, meaning parents selected to attend them. In doing so, those schools were given more resources and better training, and thus, many of the students in those schools thrived. As we began to tear down these inequities in 2017, we began to ask ourselves, "If STEM and PBL are good for some kids, why are they not good for all kids?" Consequently, we changed our approach and began to implement these tenets across our entire school district.

Across America, some schools are doing STEM, and some are doing PBL. We aimed to leverage the strengths of both. PBL and STEM can each help schools to target rigorous learning and problem solving. Given the state of our society, producing students who learn deeply and are equipped to solve problems is essential to the further positive development of our world. STEM education and PBL are not simply a process of teaching course content. They embody a process of teaching kids to be scientists, technologists, engineers, and mathematicians who are prepared to identify and address problems in our society. As we prepared to close out the 2019–2020 school year, little did we know that our STEM and PBL work would pay significant dividends as a number of very difficult moments lay in our future.

In the spring of 2020, our entire world found itself in the midst of a growing global pandemic that threatened our lives and livelihoods, and our country was at the precipice of political debate that would rip apart levels of grace and respect in ways that most of us have never seen or experienced. Our lives as school leaders would change quite drastically as we were forced to leap into a mode of thinking and planning that no one had conceived in history. Our entire world was about to make a drastic shift. After decades of exacerbating stark inequities in schools across America, we were finally confronted with a glaring need to change and to create educational spaces that truly allow for access and opportunity for all students. STEM combined with PBL assisted us in creating a playing field of relevant content, rigorous instruction, and equity. Shameful as it may appear, it took this global pandemic killing hundreds of thousands of people in the United States, to date, to unpeel the onion of inequity in America's schools and force us to realize that we truly had to change if we were committed to eradicating such constructs.

The need to change played out in real time in school districts everywhere. On Friday, March 13, 2020, the Governor of Ohio announced that he would be closing all school buildings across the state due to the growing concerns about COVID-19 and the impact it was having on communities. As cases of COVID increased, hospitalizations pushed medical facilities to the brink, and deaths continued to mount, school leaders were placed into a space where they had to plan for teaching and learning in ways that had never before been imagined. While school buildings were closing on that following Tuesday, "school" had to continue so that students could stay engaged with their learning. Teachers would have to continue to teach the curriculum. Assessment of learning still needed to take place. Special education, gifted, and language acquisition support for new immigrant students would need to continue.

The vast conundrum that we all faced at that moment was already multifaceted, and it would evolve quite quickly. It also revealed our ignorance to the lack of access that so many of our students and families had to things that many of us took for granted. How exactly would we work to create this paradigm? How would we work to organize and prepare teachers to instruct 100 percent of their students through a computer device while they all sat in the confines of their own homes? How would we approach getting computer devices in the hands of thousands of students, in a matter of days, when none of our students had been assigned and trained on this technology previously? How might we work to support parents to be able to care for their children, support their educations, provide nourishment and exercise, and ensure that they were attending school each day, all while they diligently worked to retain the jobs that enabled them to provide for their families in the first place?

Our foray into STEM and PBL would allow us to adjust more quickly than some districts, as many of our students were just as equipped to work from home as they were to work at school. As we addressed resources and infrastructure needs to provide all students with access, this perilous venture was about to take another turn that would force us to question what our curricula touted.

On May 25, 2020, video began to circulate of a Minneapolis, Minnesota street corner where George Floyd, a forty-six-year-old Black man, was killed by Minneapolis Police Officer Derek Chauvin in a heinous act of complete disregard for life, dignity, and humanity. Chauvin held his knee on Mr. Floyd's neck as he gasped for breath for nearly nine minutes, resulting in Mr. Floyd's death in the public view of the entire world. It was watched by millions across the globe and brought to light and into full reality the fears many people have harbored for years of police officers and departments in their respective localities. This incident sparked what was already a very contentious relationship between those who advocate for racial and social justice and a system of policing that many feel minimizes the lives of marginalized people.

This tragedy further complicated the need to change education. Schools, and many of the habits and beliefs taught in schools since their inception, worked to influence how our society values (or undervalues) the lives of certain groups of people. This, too, must take a shift toward something far more inclusive, and STEM and PBL can play a significant role in making that evolution come to fruition. Skills like collaboration, creativity, critical thinking, and problem solving are part of any STEM PBL and will be needed for students to learn to effectively embrace change. In a STEM PBL project, teachers teach and assess one or more of these skills. As an example, an eleventh grade Humanities course explored the following questions:

- How have federal and state legislation led to the disenfranchisement of People of Color?
- How can we teach the constitution to a younger audience so that they can be advocates for themselves and others?

A high school English/Language Arts class was tasked with the following:

- Driving Question: How does literature help readers explore and understand their world and communities in a way that enables civic involvement?
- Performance Task: You have been tasked with creating an action plan that identifies and addresses an issue within your community and, in conclusion, a product that summarizes your action plan. You will be exploring multimedia resources (readings/videos, a data website, and community agencies) to form and define your world view and, as a global citizen, what you can do to be involved in your world and create change. In your product, cite at least one reference from each resource to illustrate and clarify your perspective. Using the sources from the resource bank and your completed planning documents, create a product that summarizes your action plan. Refer to your checklist and directions for guidance.

Like the overall project, success skills are part of the glue of STEM education. Students can learn to tackle problems such as this through PBL units. This will also mean using an effective rubric for formative and summative assessment aligned to collaborating, collecting evidence, and facilitating reflection within the PBL project.

Although STEM design challenges foster this kind of assessment naturally as an organic process, PBL can add the intentionality needed to teach and assess the 21st-century skills embedded in STEM that can lead to inclusive practices and recognition of every student who walks into a classroom. All students should be able to see themselves in the curriculum. Effective use of STEM PBL can bring this type of inclusivity to reality.

Adding more to the aforementioned powder keg, our country was already embroiled in political turmoil and opposition under leadership that demonstrated wanton disregard and disdain for anyone who expressed disagreement, and this psychosis bled into the daily interactions of people. Callous disregard for others, their feelings, and their experiences, permeated the soul of America in ways that pitted sibling against sibling, parent against child, and friend against friend. These disruptions to relationships played themselves out in living rooms, at kitchen tables, in the midst of road rage incidents, in public spaces such as restaurants and grocery stores, and, most disgustingly, on social media platforms like Facebook, Twitter, and Instagram. Facebook, which was created as a platform to unite people, became a tool of opposition and divisiveness forcing us to sever ties with others whom we had known and respected since our childhoods. People began to confront people in ways that they have not done in my lifetime and to say things that expressed blatant hatred toward others who happened to be different from them. They also exercised an abundance of "keyboard courage" on their computers and cell phones by antagonizing, belittling, and insulting people that they did not know without any fear of ramifications. This would lead to discord that would play itself out in city council meetings, school board meetings, and vast other arenas where somewhat civil public discussion had been the norm.

Is STEM PBL a mechanism to address some of this incivility? Yes, it is. As mentioned earlier, collaboration, evidence collection, and thoughtful reflection are embedded in STEM and PBL. Those competencies force people of different backgrounds, beliefs, and philosophies to work together to find common ground thus forging true, connected relationships. We can use STEM and PBL to take steps to successfully overcome political division.

While navigating all of that, school leaders were confronted with the stark actuality of mimicking the school environments they were trained to create for their entire careers in a virtual environment for which they had no training whatsoever. They had to attend to academic needs, social emotional concerns, mental health and wellness, food insecurity, technology support, and the "newly discovered" equity issues that existed in districts for decades. To do so would require a new degree of creativity and innovation and a thought process that was more than simple "out of the box" thinking. It would require thinking as if there were no box.

Paving a way to overcoming such disparity requires schools to embrace a model of education that gives students opportunities to innovatively solve real-world problems, to become socially and emotionally competent, to become effective communicators who know how to collaborate, and to develop as socially aware, global citizens. Central to creating such a paradigm is the convergence of STEM and PBL and the benefits each has for closing equity gaps and creating opportunities for all students.

In addition to the integration of disciplines and attainment of success skills, voice and choice are critical components of STEM PBL. In order to ensure the engagement of all students, they each must have opportunities to shape their own learning experiences. Personal interests may motivate challenges that students want to solve, and education must embrace the notion of a passion-based methodology to teaching and learning. A second grade class was tasked with the following:

- Conservation: How can we conserve water? How does what we do affect water quality? How can we communicate the importance of the water resources to the community?
- Global Goal: Life on Land: How can we make life on land more sustainable? How can we positively affect native endangered species of Ohio?

Upper level Science students explored the following questions:

- How can we produce "near miss" stunts? (changing velocity)
- How can we predict the location of a car crash? (constant velocity motion)
- How can we imitate an Evil Knievel car through a hoop stunt? (projectile motion)
- How can we use forces to determine the mass of a rock? (force)

- How do baristas make layered drinks? (density)

When they can choose team members and products to produce to solve challenges of interest, their authentic engagement, and thus, their achievement, will increase and lead to new and different challenges for them to solve.

As educators, if we are serious about reforming education and creating a system that works for all students, we must ask ourselves a series of questions. How do we address the inequities that exist in communities where the gap between the haves and the have nots is larger than it has ever been in the history of our country? How do we assist students to develop as productive citizens armed to be change agents to improve our entire society? How do we empower students with knowledge, empathy, and preparation to be successful citizens of the world? Those questions and many others are the precursors to many conflicts and disagreements that have brought us to a place where grace, respect, and empathy are lacking commodities. We have lost our ability to disagree peaceably or to hear the perspectives that others bring. We have become isolated in our own worlds where we only listen to people who think like us and all others have ceased to be our fellow citizens, becoming our enemies instead. We find ourselves in a place of uncertainty, discomfort, untruthfulness, chaos, and confusion. We must utilize education as a means to overcome all of the things that have the potential to destroy us. STEM education and PBL can have an undeniably positive impact on all of these obstacles and help to create school environments where knowledge is applied in real-world context leading to real-world change for the better. Schools must embrace the notion that high-quality STEM education produces high-quality PBL.

The teaching of content and attention to academic disciplines are embedded within STEM; however, it is far more crucial that we focus less on the content and more on the overall pedagogical implications for effective instruction and use PBL to make learning come to life...for every student. Creating schools that see the importance of the convergence of problem-based learning and STEM education and the benefits they have to closing equity gaps and creating opportunities for all students will allow for more productive, better equipped, more empathetic citizens who care about their environments just as much as they care about themselves.

# About the Author

MELVIN J. BROWN, ED.D.

As an unapologetic advocate for public education, Dr. Melvin J. Brown has dedicated twenty-eight years of his life to creating educational spaces that are conducive to effective teaching and learning and fighting for equity in schools that were never designed for the benefit of all. He has served as a teacher, a principal in three schools, a human resources director, an associate superintendent in a large, suburban school district, a deputy superintendent, and a superintendent in an urban, suburban school district. Through it all, he has never lost sight of the need to advocate for those who cannot advocate for themselves. In this chapter, Dr. Brown explores the importance of the convergence of problem-based learning and STEM education and how they coexist to create avenues of success for students from traditionally marginalized and underserved backgrounds. Fueled by the desire to break his own cycle of generational poverty, education has paved his way to success. He firmly believes that ensuring all students receive an educational experience that provides them with educational access and opportunity will work to allow them to break cycles of poverty. Dr. Brown is Superintendent in the Reynoldsburg City Schools (Ohio). He is also a Superintendent-in-Residence and Visiting Professor at The Ohio State University in the College of Education and Human Ecology.

EIGHT

# Investing in an Equity-Focused STREAM Environment

Eudes S. Budhai

*Knowledge emerges only through invention and re-invention, through the restless, impatient, continuing, hopeful inquiry human beings pursue in the world, with the world, and with each other.* ~Paolo Freire

In the past thirty years of education, there have been extraordinary shifts based on the rapid journey of innovation. While we are learning something new, innovation continues to strengthen the capacity within education, at home and in the workplace. The critical nature of developing students at an early age for the expectations that have yet to exist is a priority in education. As a country, we are simultaneously experiencing a pandemic, social causes and issues, and an economic crisis. These matters have provided communities to have a better understanding of equity, access, and inclusion in a community.

As educational organizations, we were propelled to creating digital environments within a short timeframe and struggling with establishing an infrastructure necessary for excellence. Amazingly enough, we have come together to have conversations that would change the way we see education today. As we witness the unfolding of fluid uncertainties, there is an immediate shift to our current structures and an opportunity to be more inclusive with regard to decision making and moving towards a culture of STREAM for *all*, which includes Research and the Arts.

We understand that we are living in a time where careers are ever changing. We have a responsibility to ensure that our students are prepared and exposed to a variety of skills from an early age. Our students have been at the center of teaching and learning entering the digital world. We have explored methods to engage children in an inquiry-based model of instruction where they take ownership of their learning. It has been suggested that this model is exclusive to what is considered high achievers. Yet, we are encouraged that this is a model that is consistent with exemplary educational practices for all students. Our purpose is to generate the energy required to create 21st-century learning environments that would optimize the potential of students and professional learning opportunities.

The pandemic has shifted us to a mindset of forward thinking with equity at the forefront of discussion. We understand that all stakeholders are vital to the culmination of intensive discussion to understand the fluid nature of change. The transformation of education is indeed at our front doorsteps, and collaboration will be instrumental to the process. It is necessary to leverage the combination of external and internal experts to support forward thinking. Technology and Wi-Fi have become essential tools for teaching, learning, and maintaining effective and timely communication for all. Although there were many school districts that were ahead of the curve in providing their students with these essential tools, it was evident that the most vulnerable were struggling to receive and provide access. Thus, equity would surface to the top of the list. Districts were faced with ensuring that we sustained a culture of care, trust, and encouragement to validate our strength in collaboration, unity, and equity.

The finance and operations of districts depended on ensuring efficiencies to secure the reallocation of funds aligned to the needs of our children and professional learning. Concerned with reduction in funding sources, we became extremely frugal with our expenditures, repurposing funds, applying for funding allocation from numerous sources, and becoming proactive in forecasting budgetary constraints. Subsequently, ensuring that we were consistent with our obligations to our community and children in maximizing and leveraging funds to support our children, staff, and families.

The force of COVID-19 gave way to be mindful of the experiences, loss, and commitment to our social–emotional well-being. It was evident that self-care would be the force that propels us all to help one another. The loss of staff members, community members, relatives, and friends created a sense of urgency to engage all in mindfulness sessions of self-care. Having continuous dialogue with families led to the opportunities we are seeing today, from disruption to engagement. Family engagement turned into empowering families to authentically connect with schools. This has yielded amazing results. It has revealed that our families, despite obstacles, can understand curriculum and instruction and support our teachers with furthering the home–school connections. I have seen the power of technology through engaged families in virtual classroom activities understanding the impact of:

holding a pencil correctly, asking higher level and critical questions, note-taking aligned to listening, reading, and writing skills, and becoming experts on effectively utilizing technology resources.

Moreover, as school districts, we have practiced joining forces with our community at large. However, we know that we can all improve and develop more effective ways to be more inclusive. Nevertheless, communication became instrumental to ensure that educational organizations were consistent with transparency, addressing the needs of all stakeholders, and receiving immediate and targeted feedback. It has been important to always find a balance to accommodate sustainable and effective communication. This is the case for video conferencing, which has availed us to connect with convenience and in unconventional ways. It has provided for one-way presentations with opportunities to receive questions, two-way communications with simultaneous dialogue, screen sharing, private and general sessions, recording of sessions for future viewing, and utilizing our websites as a conduit of information as well as social media. I can recall the many sessions when our families were either excited to participate or confused about decisions being made. Either way, it was the highlight of connections, inclusivity, and transparency with our community.

This is the time to shift systems and structures as we face uncertainty and adversities. We were literally writing the blueprint while experiencing the changes within our eyes. Taking planned risks, if the term exists, was a common struggle for us all. It was the opportunity to increase the capacity of our students, staff, families, and community to have conversations about the future. All of this is occurring while we witness the transformation of working environments and the shifts to come in the next ten to twenty years. Therefore, preparing our students for this ever-changing environment will enhance their ability to be prepared. Hence, school districts will need to continue exploring new ways to establish instructional environments that empower students to solve world problems, take ownership of their learning, take risks, and learn from their experiences. Furthermore, our systems must change to accommodate for the way our children think and learn today and approach the problems that are waiting to exist. As I visit schools and see our elementary students engaged in building robots in maker space environments. The students are excited to use their creativity to create or transform robots with maker materials. Our students from kindergarten to high school are increasing their capacity through complex lessons that integrate science, technology, research, engineering, art, and mathematics. Communities inclusive of schools, partnerships, workforce, families, and students will need to connect to ensure that STREAM moves from an initiative to a movement.

This can only occur when the opportunities to engage all stakeholders exist. At the minimum, to accommodate the urgency and address the gaps that exist, it was clear that we had to tier professional learning. Seymour Papert states, "The role of the teacher is to create the conditions for invention rather than provide a ready-made knowledge." The diversity of competencies to infuse technology varies from one practitioner to another. Therefore, it was important to tier the staff into categories, build capacity within our staff, and capture the strengths of the highly competent technology innovators to serve as essential leaders. The ongoing professional dialogue and learning has resulted in our current structures, whereas in-person learning is primary with the structure of choice for synchronous simultaneous teaching and learning. We now can benefit from lessons focused on building choice boards to increase differentiation and set the stage for inquiry, exploration, and discovery.

This mindset has helped us take bold steps to ensure that we coordinate efforts and maintain a collective voice in moving forward. We understood that serving our most vulnerable students would require us to support the technology infrastructure, Wi-Fi capabilities, technology hardware, and software tools. Therefore, we develop fiscally responsive and sustainable plans for 1:1 implementation that include every child to have a technology device, Wi-Fi accessibility, and all of the resources for a high-level curriculum and shift on instruction model. This would also include the manipulatives that help students combine the creativity of design on a computer to the applications and building of structures. There is magic that occurs when students are deeply engaged and struggling with learning while collaborating.

While serving as Superintendent for the Nyack Public Schools, placing equity as the foundation for a successful district is evident. The Nyack Board of Education has been instrumental in ensuring that we continuously raise the bar on equity for all students. Having equity as a central focus made a significant difference in our daily conversations. We are in pursuit of developing new relationships, disrupting the status quo, and uncovering the talents of all students. To that extent, we are embracing a variety of topics that are thought-provoking and ensuring that all students are given the opportunities to be successful. Hence, establishing a welcoming and affirming space where the most uncomfortable conversations would be respected, encouraged, and valued. As a result, the Nyack Public Schools has approved a comprehensive equity policy outlining four pillars, our beliefs, strategies, and expected outcomes for all students.

Our students' voice plays a pivotal role in having a better understanding of how they learn best and their interest towards the future. We took the opportunity to reimagine and create a future-ready environment to increase the capacity of our digital world; environments that would facilitate teachers to engage students in a continuously changing classroom. Increasing integrating interactive technology tools, innovative furniture to increase collaboration, and making connections to your immediate community, the world and beyond. Our charge is to collaborate in teams to research and brainstorm ideas to explore ways in which we develop our students.

The power of collaboration among superintendents and colleagues from across the country made it possible to survive this thriving commitment to change. However, I had a supportive family that understood my calling, contributed to every successful opening of schools, presentations, and a continuous reminder that self-care would permit for me to continue this work. Communication tools to chat, share, and exchange ideas became the norm for many to strengthen our own capacity. This is ironic as the workforce is requiring us to prepare our students with skills aligned to the Nyack 7 Cs: critical thinking, creativity, collaboration, communication, compassion, content mastery, and cultural awareness. The commitment that the 7 Cs transcend from school to home and, ultimately, to the workforce will require deliberate and continuous attention.

# About the Author

EUDES S. BUDHAI

Mr. Eudes S. Budhai is the Superintendent of the Nyack Union Free School District, New York. He is a visionary and innovative leader with a keen sense of forward thinking to shape the lives of children, families, and communities. He has served in various capacities in the field of education for twenty-eight years.

Mr. Budhai's educational experience began in a private school as an assistant teacher working with students who required extensive emotional and academic support. Since then, he has served as a teacher, a staff developer, a department chairperson, a building and district administrator, and an adjunct professor.

Mr. Budhai serves on various local, state, and national organizations as an instructional leader advocating equitable opportunities for *all* students. He has been instrumental in developing curriculum, providing professional development to administrators, teachers, and families, and securing funds that enhance the educational experiences to reduce the equity gap for the lives of students. He supported various states adoption of the *Seal of Biliteracy* during his presidency of National Association of Bilingual Education (NABE). Mr. Budhai has been recognized by the district administrator and NYSED for sustaining graduation rates for students of color above the states average and NYSED Model School Research of Excellence.

Mr. Budhai is driven by the daily courage and perseverance exhibited by our children, families, staff, and leadership to increase our individual and collective capacity.

# NINE

# Community Recharged

ANSONIA, CT 2.0

Dr. Joseph DiBacco

Ansonia, Connecticut, like many communities across America had a rich and vibrant manufacturing history. In 1930, Connecticut had more patents issued per capita than any other state. Across the United States, the US Patent and Trademark Office issued one patent per three thousand population; Connecticut inventors were issued one patent for every 700–1,000 residents. These patents and products helped propel Connecticut into a manufacturing boom. In the early to mid-1800s, Connecticut was a leader in the manufacturing of: quality fire arms (Colt, Winchester Repeating Arms, Sturm, Ruger & Company, and Mossberg & Sons), clocks, timepieces, watches (Eli Terry, Seth Thomas, Ansonia Clock Company, Waterbury Clock Company, Bridgeport Clock Company, General Time, New England Clock Company, and Timex), metal goods (American Brass, Ansonia Copper and Brass, Waterbury Brass Company, International Silver Company), and hardware/tools (Stanley Works, Sargent Manufacturing, Yale Lock Company). In the 1920s, Connecticut became a hub of military manufacturing from shipbuilding (Electric

Boat) to helicopter (Sikorsky) and aircraft construction (Pratt and Whitney). During this time of tremendous growth and production, education and industry worked symbiotically to ensure we had a workforce for our growing economy. Public education ensured that the curriculum met the needs of the growing workforce and our thriving local economy.

In the 1980s, the need for college-educated workers increased (for high-paying jobs); the supply of college-educated workers slowed to 2 percent while the demand showed we needed a supply over 3 percent of college-educated workers to meet industry needs. Americans were led to believe that if they wanted their child to have access to high-paying jobs, college was a must. If their child and school system didn't push a pathway to college, school districts were doing the community and the child a disservice. In the 2000s, Connecticut saw high-paying jobs and large corporations leaving the state (General Electric, UBS, and Aetna). As jobs and businesses moved out or closed, many communities haven't found a way to reinvent themselves. Public education needs to be the catalyst to spawn the next Renaissance or Risorgimento. The focus of education should be forward thinking, and STEM education is linked to our educational rebirth for college or career.

## Community and Implementing STEM

Understanding my community dynamics and expectations in Ansonia, Connecticut, was imperative to me as a school leader; especially when implementing a focus on STEM education. Committing to a STEM focus has been a process, especially while trying to manage pandemic concerns. The only way systemic change can occur is if there is trust, understanding, and on-going communication with community partners. Ansonia, Connecticut, is a small, tight-knit, manufacturing community; STEM education is not a big leap. The readiness level of the community and the deep history helped support the incorporation of STEM education. Many saw STEM as the next step in our evolution as a school system and as a community.

When our community saw Ansonia High School students receive summer internships at Sikorsky/Lockheed Martin, a large aerospace and defense company, stakeholders took notice and wanted to see more partnerships. Through working with our community partner—Sikorsky/Lockheed Martin—Ansonia Public Schools was able to offer externship opportunities to Ansonia High School teachers and guidance staff. These externships were mutually designed by the district, staff, and aerospace company. To ensure we got a true picture of our students, we conducted a study to see if this interest survey coincided with our community partnerships. As a district we spent time reviewing statistics from the Department of Labor to determine what courses needed to be updated and what courses need to be reviewed.

We took a deep dive into our antiquated requirements and re-evaluated if we needed to eliminate a course offering or offer it to students as an independent study. This study was done to determine if the needs and interests of our students coincided with our new pathways and if our pathways reflected and reinforced our district's commitment to STEM education. Ansonia Public School's commitment to STEM education, community partnerships, and the need for our students to have greater options has helped us develop college and career pathways in the areas of: allied health, manufacturing, and cybersecurity. These community partnerships in STEM education have renewed students' interest in course offerings and made teachers excited to teach content that students are passionate about. The full embrace of STEM education from elementary to high school has been a community effort, the all-levels approach is our best effort to make our students more competitive in the global marketplace.

Being forward-focused is critical to the success and evolution of public education; however, this is not without its challenges. Curriculum revision is a daunting and expensive undertaking for any school district; then add the next layer of informing teachers that they need to incorporate STEM into their everyday instruction. Teacher overload is real, and we need to make STEM part of everyday instruction, not seen as one more thing we need to do. To assist in this effort, we enlist your learning community. Utilizing our private, public, and community colleges has been vital in making the STEM transformation. Districts have planned for digital learning, but having a true STEM focus district-wide takes planning, training, qualified staff, and resources. Some districts are fortunate to have staff specifically trained in STEM education while many other districts have limited funds and resources and spend their time trying to fill and retain staff; this has been a real challenge because it has become harder to recruit and retain staff during the pandemic and the national teacher shortage.

Understanding the limitations of staff and available resources, school leaders look toward technology and on-line resources to help buttress STEM education. The teacher shortage cannot prevent school districts from finding ways to give students the best opportunity for success. Looking at how other communities have partnered with on-line STEM programming has made a considerable difference in providing high-quality STEM education. Districts often decide which students will receive the benefits of STEM education. Our district made a promise to provide STEM to all learners, and we made sure that most underrepresented students in STEM education (low-income, female, and minority students) had the same level of access.

## A Call to Action

The day I heard Barak Obama tell Congress and the nation that STEM education, "is our generation's Sputnik movement" was a call to action. I was compelled to do more with STEM education because I kept reading about companies that were moving out of state or out of the country because they wanted to be closer to their workforce. It was clear to me that we needed to step up our game in education and work with our business partners to ensure that companies, skilled jobs, and career opportunities remained part of the local community. The approach we have taken to STEM education is to make it available to all students, dispel STEM stereotypes, and make STEM learning available at all hours (on-line) and accessible to our afterschool partners as well. We are by no means where I would like to see our district in STEM education, but we are much further along than I could have ever dreamed. It is a hope of mine that rather than seeing STEM as separate content areas, we will be able to seamlessly blend content in a manner consistent with STEM education and students will build the skills and abilities to be a force in the global marketplace.

# About the Author

DR. JOSEPH DIBACCO

Ansonia 2.0—creating a rebirth in a "manufacturing belt" community is what Superintendent of Schools Dr. Joseph DiBacco is doing in Ansonia, Connecticut. DiBacco's district has been highlighted multiple times on WTNH "What's Right With Schools" highlighting a middle school woman in engineering partnership with Sikorsky and recently showcasing Ansonia High School's STEM career pathways in engineering, allied health, and cyber-security. DiBacco's work with the Teamsters and Sikorsky has helped create summer engineering internship opportunities for high school students. Dr. DiBacco's passion is infectious, and it wasn't a surprise that he was asked to deliver the commencement speech at Quinnipiac University School of Education, Class of 2021, on May 9, 2021. Dr. DiBacco is a leader dedicated to creating opportunities and fostering community and business connections that will allow his students to have opportunities they never saw possible.

# TEN

# The Importance of STEM and Early Access

## HOW SALT LAKE CITY SCHOOL DISTRICT IS MOVING THE NEEDLE

Timothy Gadson III, Ph.D., Candace Penrod, M.Ed., M.S., and Frederico L. Rowe, Ed.S.

STEM education is essential to producing a STEM-literate citizenry and creating authentic connections and contexts for educating America's youth (Holmlund, Lesseig & Slavit, 2018). Today, many local and global challenges require STEM-minded teams to work collaboratively to find and solve complex problems. Harlen (2010) suggests science education should enable every individual to take an informed part in decisions and appropriate actions that affect their well-being and the welfare of society and the environment. In K–12 education, partners, funders, and community organizations often seek to bolster and expand STEM programming in secondary schools to prepare more STEM-interested and STEM-prepared high school graduates (Allen, Lewis-Warner & Noam, 2020; National Research Council, 2011; Ralston, Hieb & Rivoli, 2013). However, suppose students are not exposed to opportunities and coursework well before reaching high school. In this case, their interest and capacity to select or excel in STEM-related coursework are often unattainable.

Intentional and well-designed exposure to STEM is essential in early learning (DeJarnette, 2012). Students in early childhood and early elementary education need the opportunities to engage in STEM. STEM education benefits students' thinking, curiosity, creativity, communication, and problem-solving capacities (Klimaitis & Mullen, 2020; Rosicka, 2016; Turiman, Omar, Daud, & Osman, 2012). STEM benefits society by educating an innovative citizenry that can tackle difficult and complex problems and fostering individuals who can engage in STEM fields and careers. Early access to STEM ensures all students will become aware of STEM opportunities available to them throughout their PK–12 educational experiences and beyond (DeJarnette, 2012). All children are born curious with a great capacity to learn and develop critical thinking skills, attain STEM content knowledge, and build a positive self-identity toward STEM.

Waiting to expose students to science, computers, or technology until upper elementary grades or beyond does not provide equitable access to STEM (Jackson, et al., 2021). This practice also prohibits the growth in STEM knowledge and perpetuates perceptions that STEM is not for everyone. It is time to elevate STEM in early childhood and early elementary programming for all students so that STEM pathways will become viable options for all students.

## Challenges and Progress in Early STEM Access

Early access excitement was in the air! A loud buzz filled a Title I first-grade classroom in Salt Lake City School District as a diverse classroom of students in the urban school district experimented with syringes, stoppers, and colored water. A group of three girls excitedly "figured out" how to make water move through a tube and shoot out of the syringe like a fountain and then started to explain what was happening. "Science is so cool!" could be heard across the room as another student tested his ideas to see what he could figure out. These early learners were engaged and enthralled in science and engineering. All of this was made possible thanks to a generous donor who provided the materials, a district vision that all students can learn and should have an equitable opportunity to learn, and a supportive education foundation. Early STEM learning matters. It mattered to the twenty students in that classroom that day!

Providing early access to STEM includes intentional consideration of culture, materials, and teacher education. All children enter educational environments with a family and community culture. This culture, including ways of knowing, being, and doing, is essential for each student (Medin & Bang, 2013). STEM education should therefore build a collective classroom community culture. This culture building should consider the multiple perspectives of students and their families, which requires intentional conversation and interaction whereby students and their parents are considered active partners in the learning experience. Community circles, classroom chats, open houses, parent nights, home visits, focus groups, and surveys are all excellent ways to welcome voice and, in turn, empower students and parents to help shape the experience that students receive in our classrooms (Zepeda, 2019).

Challenges to providing early access to STEM education should be identified and addressed. One significant barrier to delivering quality STEM education in the primary grades is that formal support for and high expectations concerning science teaching and learning is less of a priority than teaching literacy and mathematics at the local, state, and national levels in our elementary classrooms (Smetana & Coleman, 2015). Other barriers include a lack of funding for materials and supplies, training for early childhood teachers in STEM practices and mindsets, and competing priorities for time and resources. Further, professional learning opportunities through instructional coaching and other content area experts are less likely to be provided outside the focus on literacy and numeracy (Spillane & Hopkins, 2013). As mentioned earlier, another significant barrier is staffing. Teachers at the early childhood and elementary levels have varying degrees of experience, interest, and comfort with teaching science and other STEM-focused content. Early childhood and elementary teachers are typically trained and considered subject area generalists; therefore, they may feel unprepared to facilitate science learning (Fulp, 2002). Some early childhood and elementary teachers' hesitation to teach science could create an uninviting atmosphere for community involvement, thus, hindering early access to STEM learning and engagement.

Nevertheless, progress has been noted when stakeholders develop partnerships that provide expertise, financial, and human capital from business and industry. This progress is expanded when educational entities collaborate with STEM-rich institutions, such as local gardens, museums, and zoos, to provide educational programming and experiences for both students and families. Some communities have also found success through offering transportation to families across town to participate and learn from activities.

It is important to note that in the past, STEM efforts have lacked concern for the culture and context of many students, resulting in a severe chasm between STEM programs and STEM participation for underrepresented students (Bang, et al., 2017; Elrod & Kezar, 2017; Jones, et al., 2018). Hence, everyone wins when culture matters and early access to STEM is approached from multiple viewpoints.

Educators must understand the importance of bolstering STEM pathways early on. Many students and communities of color are underrepresented in STEM programs in institutions of higher learning and STEM careers (Jones et al., 2018). STEM pathways to careers, higher education, and informed citizenry start in early childhood and continue through a student's K–12 experience. All students deserve access to high-quality STEM experiences from the beginning of this educational journey. Society needs their ideas, brilliance, and innovation, and they need our support.

## Solutions to Early Access in STEM

As a result of federal mandates, students' learning and engagement in subject areas other than numeracy and literacy are highly dependent upon teachers' comfort and content knowledge and the time allocated for other subject areas in the daily schedule of learning blocks (Spillane & Hopkins, 2013; Supovitz, Sirindes & May, 2010). Access to training and materials on how to teach science, engineering, and technology is a necessary first step in removing barriers to STEM education in early childhood and elementary programs. Integrated learning experiences for young children that weave together literacy and STEM is the next step in ensuring adequate time for teaching STEM-focused content (Kelley & Knowles, 2016).

It is time to focus on access to early STEM education in PK-12 learning systems (NRC, 2021). We need to work collaboratively with our diverse communities and invite all stakeholders and voices to the table to find ways to increase equitable access to early STEM learning and engagement for all, especially youth in groups currently underrepresented in STEM careers. Many organizations have programming and outreach opportunities to share with PK–12 schools and districts (Allen, Lewis-Warner & Noam, 2020; Berry, 2020; Boulden, et al., 2020). These opportunities will expand teachers' STEM knowledge and ability to provide sound and consistent STEM learning experiences for all students that are appropriately rigorous and highly engaging.

In schools an entire segment of the K–12 population is underrepresented in STEM education (Prescod, et al. 2020). Correspondingly, teachers must be intentional about engaging students of color and providing programs that promote learning in culturally appropriate ways. The use of lesson studies, professional learning communities, and instructional coaching are tools that could be leveraged to increase teacher capacity around engaging students of color in the learning (Zepeda, 2019; Hinnant-Crawford, 2020). For example, honor the tremendous STEM knowledge in science and engineering tied to the clay work of Indigenous people (Barajas-Lopez & Bang, 2018). Students must learn about and see the contributions of people who look like them in the lessons they participate in and are engaged in (Allan, 2016; Thompson, et al., 2016).

PK–12 educators must partner with businesses to learn and understand industry standards that should guide the appropriate infusion of technology in the learning experiences provided in classrooms. Put measurement and investigative devices in young students' hands. Engage young learners in digital and technology-based learning and thinking. Provide experiences both in and out of school. Create maker spaces to allow students to tinker, explore, and create (Hachey, An, & Golding, 2021). Encourage local libraries and businesses to support STEM initiatives and activities. Work with local universities in partnerships to provide professional learning for teachers of early childhood and elementary STEM subjects. One such program is the University of Utah's Elementary STEM Endorsement Program, which collaborates with local Utah school districts to increase elementary teachers' capacity in science and engineering, technology integration, and mathematical thinking. Further, educators should foster relationships with university scientists and engineers to work in local classrooms to support teachers and students engage in STEM experiences.

Educators should create a district or state vision that includes a STEM continuum that supports all learners. Stakeholders representative of all constituencies should be invited to develop this vision. There should be a deliberate focus on exposure to STEM in early childhood and K–2 educational programming. Research has provided evidence that young children possess a broad repertoire of cognitive capacities enabling them to engage in science from a very young age (NRC, 2007). Most importantly, these capabilities related to the scientific endeavor do not develop on their own, apart from instruction, but, rather, they grow under the guidance of a scientifically literate individual (NRC, 2007). Therefore, these capacities must be encouraged, nurtured, and sustained in supportive science learning environments throughout a child's PK–8 experience (NRC, 2007). Early access is critical.

Connect STEM curriculum departments with career and technical education to build strength and coherence of PK–12 programming. Create internships and apprenticeships for high school students using backward design, starting with what high school students should learn from internships and apprenticeships. Then plan the preparation needed at each learning level, including what students should know and be able to do in early childhood and elementary programs that will ultimately allow them to apply their learning in more mature and complex learning and work environments. It is essential to realize that early learners today will be high school students tomorrow; therefore, they need access to STEM learning opportunities now.

The Salt Lake City School District's vision for STEM education is grounded in engaging vital stakeholders and creating innovative opportunities to engage students in STEM. PK–12 STEM education can be viewed as a process of the iterative and creative nature of the enactment of each discipline in which the learner uses thinking, innovation, problem-finding, and problem-solving strategies to engage in meaningful content-related interactions (Executive Office of the President, 2018; National Research Council, 2011; Hess, et al., 2011). Our vision for STEM includes both individuals and programs in public and private sectors, including educational programs, research and development activities, internships, apprenticeship, and preparing all students to become STEM-literate by high school graduation.

## Our Vision Evolves

While self-bagging at a local grocery store, a young girl noticed my STEM T-shirt and said with enthusiasm in her eyes, "Oh, I did STEM once, and I really liked it!" I was thrilled that she shared this with me but was stunned that she used the word "once." Her interest and enthusiasm for STEM were apparent, as was the reality that she might not have other STEM opportunities if left to find them herself. At that moment, the realization hit that many of our students are relying on us, as educational leaders, to build accessible bridges to STEM education and rigorous learning opportunities. This young lady was totally dependent on our educational system to provide her with access to STEM programming and experiences, and I wasn't confident we would deliver. Experiences like the one described here have been a driving force to ramp up previous efforts to ensure early access to STEM education in Utah.

The Salt Lake School District Department of Science began working with the Salt Lake Education Foundation and local donors to align STEM funding to address our students' greatest needs. This work exposed considerable gaps in STEM materials and curriculum in early childhood and early elementary classrooms and teachers' capacity to provide hands-on learning in science and engineering. Through the collaboration, a local STEM-based business provided funding to train teachers and purchase materials for students. District staff provided classroom support to build early childhood and elementary teachers' confidence in teaching STEM-related content. There was power in aligning core educational standards to address teaching and learning needs and provide resources to create robust and sustainable STEM programming. Alas, equitable access to early STEM education was unfolding.

District staff looked for ways to scale up successes through business partnerships, the Salt Lake Education Foundation, and the school district to increase the positive impact on teachers and students. As staff worked closely with national STEM educators, it became apparent that what was working in other areas and states similar to Utah was working in the Salt Lake City School District. District leaders realized that strong STEM communities have a strong STEM infrastructure connecting individuals and programs in more intentional and profound ways. Staff learned about the STEM Learning Ecosystem Community of Practice and quickly related to their collective vision of how an ecosystem approach to STEM bolsters support for the most vulnerable learners along a PK–16 continuum. With the STEM Ecosystem Community of Practice's support, the Utah STEM Ecosystem was created. Members of the Ecosystem met with the greater STEM community across Utah to streamline a STEM vision, purpose, and resources and identify areas of greatest need. Community members were inspired by the work already underway across Utah. They welcomed opportunities to identify gaps, particularly those in early STEM access, to begin taking steps to increase access and equity of opportunity for all students. Working together across early childhood, science, mathematics, digital learning, and career and technical education, community members are strengthening early STEM programming and inviting innovation such as working on an English/Language Arts (ELA) aligned science and engineering education model for early childhood. This includes place-based learning in K–6 classrooms and implementing an early mathematics curriculum aligned with core standards and curriculum used in elementary classrooms. Community members are investigating how to bring computer science into early childhood and elementary classrooms to open opportunities for students in STEM fields. Indeed, it takes a village to ensure access to early STEM.

## Concluding Remarks

In 2010, during a press release on the Change the Equation Initiative, President Barack Obama asserted that "our nation's success depends on strengthening America's role as the world's engine of discovery and innovation" (The White House, Office of the Press Secretary, 2010). President Obama continued by emphasizing the need to educate students today, "especially in science, technology, engineering, and math," to provide tomorrow's leadership (p. 7 ). His message is still relevant today, with even greater urgency. Early access to STEM lays the foundation for this vision to become a reality. It is up to us all as educational leaders, business and industry stakeholders, teachers, policy-makers, institutions of higher learning, STEM-rich institutions, and others to come together and support STEM programming for our youngest learners. STEM should never be a one-time experience for our students. Early access to STEM education must start now.

## References

Allan, C. (2016). Windows and mirrors: Addressing cultural diversity in our classrooms. Practical Literacy: The Early and Primary Years, 21(2), 4–6.

Allen, P. J., Lewis-Warner, K., & Noam, G. G. (2020). Partnerships to Transform STEM Learning: A Case Study of a STEM Learning Ecosystem. Afterschool Matters, 31, 30–41.

Bang, M., Brown, B., Calabrese Barton, A., Rosebery, A., & Warren, B. (2017). Toward more equitable learning in science. In Schwarz. C. V., Passmore, C., & Reiser, B. J. *Helping students make sense of the world using next generation science and engineering practices* (pp. 33–58). NSTA Press.

Barajas-Lopez, F., & Bang, M. (2018). *Indigenous making and sharing: Claywork in an indigenous STEAM program.* Equity & Excellence in Education, 51(1), 7–20.

Berry, T. (2020). Greater Austin STEM Ecosystem. In STEM in the Technopolis: The Power of STEM Education in Regional Technology Policy (pp. 171–188). Springer, Cham.

Boulden, D., Edwards, C., Cateté, V., Lytle, N., Barnes, T., Wiebe, E. N., & Frye, D. (2020, March). Creating a School-wide CS/CT-focused STEM Ecosystem to Address Access Barriers. In 2020 Research on Equity and Sustained Participation in Engineering, Computing, and Technology (RESPECT) (Vol. 1, pp. 1–2). IEEE.

DeJarnette, N. (2012). America's children: Providing early exposure to STEM (science, technology, engineering and math) initiatives. Education, 133(1), 77–84.

Elrod, S. & Kezar, A. (2017). Increasing student success in STEM: Summary of a guide to systemic institutional change. Change: The Magazine of Higher Learning, 49(4), 26–34.

Executive Office of the President. (2018). *Charting a course for success: America's strategy for STEM education.*

Fulp, S. L. (2002). *The 2000 national survey of science and mathematics education: Status of elementary school science teaching.* Chapel Hill, NC: Horizon Research.

Hachey, A. C., An, S. A., & Golding, D. E. (2021). Nurturing Kindergarteners' early STEM academic identity through makerspace pedagogy. Early Childhood Education Journal, 1–11.

Harlen, W. (Ed.). (2010). *Principles and big ideas of science education.* Hatfield, England: Association for Science Education.

Hess, F. M., Kelly, A. P., & Meeks, O. (2011). *The case for being bold: A new agenda for business in improving STEM education.* Institute for a Competitive Workforce.

Hinnant-Crawford, B. (2020). Improvement science in education: A primer. Myers Education Press.

Holmlund, T. D., Lesseig, K., & Slavit, D. (2018). Making sense of "STEM education" in K-12 contexts. International journal of STEM education, 5(1), 1–18.

Jackson, C., Mohr-Schroeder, M. J., Bush, S. B., Maiorca, C., Roberts, T., Yost, C., & Fowler, A. (2021). Equity-Oriented Conceptual Framework for K-12 STEM literacy. International Journal of STEM Education, 8(1), 1–16.

Jones, J., Williams, A., Whitaker, S., Yingling, S., Inkelas, K., & Gates, J. (2018). *Call to action: data, diversity, and STEM education.* Change: The Magazine of Higher Learning, 50, 40–47.

Kelley, T. R., & Knowles, J. G. (2016). A conceptual framework for integrated STEM education. International Journal of STEM education, 3(1), 1–11.

Klimaitis, C. C., & Mullen, C. A. (2020). *Access and Barriers to Science, Technology, Engineering, and Mathematics (STEM) Education for K–12 Students with Disabilities and Females.* Handbook of Social Justice Interventions in Education, 1–24.

Medin, D. L., & Bang, M. (2013). *Culture in the classroom.* Phi Delta Kappan, 95(4), 64–67.

National Research Council. (2011). Successful K-12 STEM education: Identifying effective approaches in science, technology, engineering, and mathematics. National Academies Press.

National Research Council. (2021). *Science and engineering in preschool through elementary grades: The brilliance of children and the strengths of educators.* Washington, DC: The National Academies Press.

Prescod, D. J., Haynes-Thoby, L., Belser, C. T., & Nadermann, K. (2020). Including Gottfredson's Career Theory in STEM Initiatives Geared toward Students of Color. Journal of Negro Education, 89(2), 158–168.

Ralston, P. A., Hieb, J. L., & Rivoli, G. (2013). Partnerships and experience in building STEM pipelines. Journal of Professional Issues in Engineering Education and Practice, 139(2), 156-162.

Rosicka, C. (2016). *Translating STEM education research into practice.*

Smetana, L. K., & Coleman, E. R. (2015). *School Science Capacity: A Study of Four Urban Catholic Grade Schools.* Journal of Catholic Education, 19(1), 93–128. https://doi-org.pallas2.tcl.sc.edu/10.15365/joce.1901192015

Spillane, J. P., & Hopkins, M. (2013). *Organizing for instruction in education systems and organizations: How the school subject matters.* Journal of Curriculum Studies, 54(6), 721–747. http://dx.doi.org/10.1080/00220272.2013.810783

Supovitz, J., Sirinides, P., & May, H. (2010). *How principals and peers influence teaching and learning.* Educational Administration Quarterly, 46(1), 31–56. http://dx.doi.org/10.1177/1094670509353043

The White House, Office of the Press Secretary. (2010, September 16). *Remarks by the President at the announcement of the 'change the equation' initiative.* Available at www.whitehouse.gov/the-press-office/2010/09/16/remarks-president-announcement-change-equation-initiative

Thompson, C. M., Catapano, S., Carrillo, S. R., Fleming, J. (2016). More Mirrors in the Classroom: Using Urban Children's Literature to Increase Literacy. United States: Rowman & Littlefield Publishers.

Turiman, P., Omar, J., Daud, A. M., & Osman, K. (2012). *Fostering the 21st century skills through scientific literacy and science process skills.* Procedia-Social and Behavioral Sciences, 59, 110–116.

Zepada, S. (2019). Professional development: What works (3rd ed.). Routledge.

# About the Author

TIMOTHY GADSON III, PH.D.

A staunch advocate for equitable access to learning for all students, Timothy Gadson is serious about engaging in the national conversation to increase the participation of students of color and females in STEM. Timothy is superintendent of the Salt Lake City School District with twenty-eight years of experience. Timothy believes that all children can learn and do learn when the curricular content and instructional delivery are quality, culturally responsive, and personally relevant. He notes the importance of adults building strong, trusting, and lasting relationships with all students daily to support their learning and reduce the impact of trauma. More specifically, Timothy knows it takes an effective teacher in every classroom, every year, delivering quality, rigorous, relevant, and data-informed instruction to ensure every child graduates as a global citizen ready for college, career, and life.

In his chapter, Timothy stresses the need for early access to STEM education and offers insight into what is currently being done in the Salt Lake City School District to move toward this aim. Timothy received his bachelor's degree in Business Economics and Secondary Education from Florida Agricultural and Mechanical University in Tallahassee, Florida. He earned his master's and doctor of philosophy degrees from Washington State University in Pullman, Washington.

# About the Author

## CANDACE PENROD, M.ED., M.S.

A true advocate for science education, Candace Penrod brings passion, insight, and over twenty years of experience to the conversation. Candace believes that access for all means all and works tirelessly as the Salt Lake City School District Science Supervisor to ensure this vision for all students. As a STEM advocate, she is the co-founder of the Utah STEM Ecosystem Community of Practice.

In her chapter, Candace outlines a robust argument for early access to STEM and offers experience in leading change in this urgent matter. Candace received a B.S. in Elementary Education from Brigham Young University, an M.Ed. in Teaching, Learning, and Literacy from the University of Utah and an M.S. in Earth Science Teaching, also from the University of Utah. She is currently a doctoral student in education at Utah State University with an emphasis on science education.

Candace has a love for all things geology and enjoys spending time in Utah's many National Parks. She has been known to say that "Science ROCKS!"

# About the Author

FREDERICO L. ROWE, ED.S.

As an elementary school principal, Frederico L. Rowe has seen firsthand the lack of support for quality science education in early childhood and elementary programs. Frederico supports quality professional learning through instructional coaching, professional learning communities, and a focus on the cultural capital students bring to the classroom every day. While serving as principal in Atlanta, Georgia, Frederico partnered with staff at Georgia Tech to bring coding to students in underserved and underprepared communities. He believes that students, regardless of level, exceptionality, and background, deserve the opportunity to experience quality STEM education.

In his chapter, Frederico emphasizes the importance of teacher development, materials and resources, and a supportive environment for STEM education that can stand alone or be part of an interdisciplinary approach to learning. Frederico received his bachelor's degree from Alabama A&M University in Elementary/Early Childhood Education. He received advanced degrees from Georgia State University in Middle Grades Mathematics and Informational Technology and Columbia University's Teachers College in Organizational

Leadership. Frederico has a Specialist Degree in Educational Leadership and is pursuing a doctoral degree in Curriculum Studies from the University of South Carolina.

# ELEVEN

# Elementary STEM Building Blocks

## A STEP-BY-STEP APPROACH FOR EDUCATIONAL LEADERS

Judith A. LaRocca, Ed.D.

Today's educational leaders may find that preparing students for the future presents an ill-defined challenge that often feels like a moving target. While none of us have a resident crystal ball, administrators who are attending to the needs of workforce development know that a solid foundation in STEM will be critical for opening the door to lifelong opportunity for students. This linchpin work cannot wait and must start with innovative leadership at the elementary level.

Leading a prekindergarten through sixth grade school district, prioritizing foundations in science, technology, engineering, and mathematics should be foundational to our leadership practice. It can be easy to lose sight of this work as the focus in elementary education often leans heavily on literacy instruction. Teaching children to read is a priority goal, particularly in the primary grades. Nonetheless, educational leaders also have a moral obligation to begin preparing our students *now* for the future they will inherit at the end of their educational journey.

That groundwork is incomplete without a robust education in STEM concepts and practices.

How can elementary leaders ensure students are ready for an unpredictable tomorrow, and how do we move forward? With so many competing primacies, it is easy to feel overwhelmed or stuck by inaction. Principals and curriculum experts may not know where to begin when faced with the wide-ranging task of implementing comprehensive STEM education at the elementary level. Teachers need professional learning to support and meaningfully change their practice, and budgets are often at capacity, making it difficult to align resources to STEM goals. However, it is possible and urgent. As a country and public institution, we're already behind.

The way forward is through taking a step-by-step approach, building blocks, that lead to increasing district-wide capacity for STEM education. Effective implementation strategies utilize an incremental approach to build capacity carefully and intentionally over time. This constructive approach is also budget friendly. Don't let the mentality of scarcity stop you from achieving your STEM mission. Find an entry point, and move forward one step at a time.

Wherever you look, STEM can be found; it is everywhere in our world. Taking a step forward, however small, will bring you further than you were before. As educational leaders, the stakes are high to make the best decisions for our children. We can become paralyzed with fear and worry that we won't get it *right* for our kids. This fear can stop us in our tracks and prevent us from acting. When this happens, remind yourself that there is more than one way to get it *right*. Move past the worry and fear, and push forward for your students.

While research-based practice delivers positive results, leaders must also recognize that context matters. Our districts and schools are unique, and these strengths enable us to find the approach that works for our students, our teachers, and our families. We should remind ourselves that when it comes to innovation, we may find ourselves in unchartered waters. Use building blocks to smooth the waters and deliver results. Before you know it, you'll be graduating students who possess the real-life skills to thrive in a future they helped design for themselves.

The building block approach includes several core areas of focus. First is the implementation of a guaranteed and viable curriculum paired with comprehensive teacher professional development. The combination of a robust local curriculum and effective instructional practice in place, leaders can begin introducing innovation and later explore expanded partnerships. Your STEM initiative progress may include these same building blocks; however, leaders should consider performing a needs assessment and subsequently identify building blocks that can meet the unique needs of their individual district and schools.

## Building Block #1 - Curriculum and Instruction

A guaranteed and viable core curriculum[1] in science and mathematics is often overlooked in the pursuit of innovative ideas in STEM education. The myriad of mandated content areas, combined with the emphasis on literacy at the elementary level, can overshadow strong content and pedagogy in mathematics and science in elementary schools. Adding to the challenge are ongoing changes to national and state learning standards. And worse, with limited funding, districts may adopt new programs without a hardy professional development plan, leaving teachers unprepared and afraid.

Districts should leverage existing systems to carry the implementation load. Internal curriculum committees can conduct a review of their local science and math programs. The goal of the review is to identify any gaps in the local curriculum to ensure that curriculum resources are aligned with current standards and assessments. New science programs should align to the Next Generation Science Standards[2] (NGSS) as they include not just Earth, Physical, and Life Science, but also Engineering. Additionally, the NGSS identify practice in Crosscutting Concepts such as "cause and effect" along with Science and Engineering Practices, all attributes that support a well-rounded STEM initiative. Standards-aligned math programs that support fluency, along with number sense and deep conceptual understanding, will also be supportive. Look for programs that have the eight Standards for Mathematical Practice[3] built in.

Unlike their secondary school counterparts, elementary teachers do not specialize in specific content areas, leaving gaps in their professional learning. Yet many school leaders and the very nature of their job set the expectation for elementary teachers to be experts in every area. Targeted professional learning is significantly impactful in ensuring that elementary teachers are adept at helping students develop deep conceptual understandings in science and math. Ensure the fidelity of any program implementation by providing teachers with multiple opportunities for stable and focused essential learning during the first and second years. Be sure to include principals and other school leaders to enable their leadership and support effective teacher practice through the observation process. Professional development must not be an afterthought to a program adoption. It is imperative that resources are aligned to the STEM goal, and that includes investing not just in program but also the people who are expected to deliver results.

## Building Block #2 - Innovation

Innovation doesn't need to cost a lot, and it doesn't have to start all at once. Incrementally building your work in this area can yield excellent results. Five years ago, one of the first things our district started were STEM-related projects that teachers could easily implement in several science periods. At the time, the district had an outdated science program without enough textbooks for every student. Teachers were doing their best to meet the NYS standards for science, but the next curriculum adoption in our budget cycle was allocated for a math program; thereby, an update for science instruction was not on the docket.

The solution started with a summer curriculum project where teachers collaborated to create one STEM project per grade with a simple blackline master copy and a short list of consumable supplies, many of which were already on hand. These hands-on STEM projects included activities such as the marshmallow tower challenge, engineering a mountain rescue, moon lander design, and catapult challenge. The next summer the initiative expanded by adding a second project to each grade. That same year the maker space concept was introduced to students by several teachers who had a passion for this idea. Most of the startup maker spaces began with existing art supplies. In year three, the maker space idea was expanded, and each building housed an Innovation Lab that replaced outdated computer labs. Every building was allocated several hundred dollars to buy STEM-related materials, such as KEVA® Planks, Breakout EDU, and Bee-Bot®. This original concept had teachers bringing students to the Innovation Lab on a first-come first-served basis—a great start, but it still needed much work.

To fully build STEM capacity across the district, we had to create a new system intentionally aligned to that goal. The director of instructional technology and innovation, along with district instructional coaches, led the way in operationalizing the STEM initiative by creating a local STEM curriculum that leveraged the Innovation Labs in each school as a resource. The curriculum was based on the Picture-Perfect STEM[4] books from the National Science Teaching Association (NSTA). These books cleverly pair picture books with STEM concepts. Now, teachers are scheduled to visit the Innovation Labs with their class twice a year for two weeks at a time. This allows teachers to fully immerse students in a STEM environment for hands-on lab activities. Teachers are also encouraged to visit the Innovation Lab at other times throughout the year as availability allows. There are two units each for grades K–6, and they include topics such as Withstanding Weather–Building Robust Structures, Bionic Animals, and Cleaning the Oceans, all using Lego® WeDo.

The Innovation Lab, with its STEM curriculum and resources, is one of the most successful district STEM implementations of the past five years. The concept of the Innovation Lab was expanded over time and grew out of a commitment to providing opportunities for innovation to our students. This commitment also saw the creation of a district-wide fourth grade engineering project, a new NGSS-aligned science program, a district-wide online coding program, after-school STEM clubs, and a new STEM content-driven learning management system. Collectively, resources were aligned to the goal of providing students with multiple opportunities to engage in and build their foundational STEM knowledge.

## Building Block #3 - Partnership

One of the valuable building blocks on our STEM journey has been leveraging partnerships with outside organizations. Partnerships are an essential tool to advance opportunities that enrich STEM experiences for students and families. They provide diverse perspectives and divergent complexity that can energize your team and your students.

Being located very close to the biggest metropolitan center in the United States, many of our students have little awareness of where their food comes from. A small farm located close to our community provides an opportunity to give our students access to real-world science close to home. Students participate in a series of farm visits, learn how to prepare soil and plant seeds, care for their plants, and ultimately harvest and eat what they have grown. Students have multiple interactions with farm volunteers throughout the growing season and engage in inquiry and writing activities throughout the experience. Families can visit the farm where student plants are growing and have access to expanding their understanding of the farm-to-table process. This year will be the district's sixth year on the farm, with over 1,500 students impacted through this valuable partner experience.

In continued pursuit of providing students with STEM foundations at the elementary level, we formed a partnership with TechTrep™[5]. Their mission to future-proof kids for a 21st century STEM-based workforce is in line with our community's wish for students to show achievement in STEM.[6] The partnership provides for a content-rich online learning system that includes activities on topics ranging from drawing and animation to entrepreneurship to computer programming. While there are many online resources with related content, the advantage is in being able to customize the courses to work within our systems with a focus on advancing STEM at the elementary level. We plan to work with TechTrep™ over the next five years to build a progressive course sequence that will expand one grade level each year. This model will work within our existing library media program and provide all students with multiple opportunities for STEM enrichment. The Tech-Trep™ partnership is the next phase of our plan for ongoing STEM initiatives.

In New York, all public-school districts have existing partnerships with our local Board of Cooperative Educational Services (BOCES). Through our local BOCES, teachers are participating in a state-funded Smart Start Grant[7] whose purpose is to develop, implement, and share innovative programs that provide professional development and support to increase expertise in educational technology among teachers in grades K–8. With BOCES as the grant coordinator and facilitator, multiple districts are working with New York Institute of Technology (NYIT), providing no cost graduate course work and regular professional development to small teams of teachers. In addition, one elementary school in the district has an established direct partnership with NYIT that has provided professional development along with on-site visits of their mobile STEAMed Van for hands-on STEM experiences.

Our district has also leveraged other partnerships over the past five years. The Valley Stream community has an educational foundation that affords financial and creative supports for educational initiatives. District administration partners with the foundation to advance district goals, and the foundation has provided resources to the Innovation Labs, such as 3-D printers. The local interschool parent–teacher association provides annual mini grants of several hundred dollars to teachers. These grants are often STEM related and have included funding for hands-on building and design materials.

There are multiple opportunities to create comprehensive experiences for elementary students in STEM that simultaneously ground your step-by-step approach. You can grow your efforts utilizing the aforementioned building blocks that include highly effective curriculum and instruction, innovative projects, and expansive partnerships. Elementary school leaders must commit to graduating students with real-life skills they need for future success. All of us know that a successful high school graduate is K–12 effort. In 2016, our school community recognized the importance of STEM, identifying it as one of five goals in its strategic plan. The focused efforts of district leadership, along with effective teaching and learning, has resulted in our elementary students having a well-rounded education that is future proof. Find your building blocks for STEM foundations and your district will get it *right* for students.

# About the Author

## JUDITH A. LAROCCA, ED.D.

Dr. Judith LaRocca is an educational leader known for her work in the areas of inclusive hiring practices and innovative STEM instruction. In Dr. LaRocca's chapter, she shares practical strategies for implementing solid foundations of STEM that can be built over time to ensure students' future success. Featured in The Education Trust's *Educator Diversity Playbook* (https://seeourtruth-ny.edtrust.org/playbook/), Judith highlights ways to support diverse teacher candidates and eliminate barriers in the hiring process. Over the past fifteen years, she has effectively led this work in two K–6 school districts in New York. Judith lives on Long Island, New York, with her husband Joel and son Eric, and has a passion for making quilts.

# TWELVE

# Making of a STEM Town

Dr. Salvatore Menzo

Over my last eighteen years as a superintendent in Connecticut, I have witnessed a transformation in the relationship between business and education. At first, it was a slow awareness of each other, but now it has become a virtual requirement for such partnerships to be established and flourish in a community. While Superintendent of the Wallingford Public School District, I was blessed with the opportunity to collaborate with incredible businesses, community partners, and my fellow educational staff, students, and families to develop something that truly transformed the entire town.

## Connected Needs

Every school district that receives Carl D. Perkins Career and Technical Education funding must establish a community advisory group to ensure that the district is establishing programming that best meets the needs of students and maintains relevance in the business community. This committee can be one of the single most powerful groups if a superintendent chooses to leverage them in such a manner. I was fortunate to assemble a great group of business, community, and educational leaders in Wallingford who were committed to improving education while we were driven to ensure the growth of the town. This symbiotic connection was essential in fostering a relationship resulting in mutual benefit for the economic wellbeing of the town and educational opportunities for its children.

In the many meetings held each year, the need to develop a STEM-minded workforce was clear. Wallingford, Connecticut, being the number one per capita international exporter of goods in the United States, required trained personnel to fill new positions as well as those to be vacated by a veteran workforce. Without these positions filled, companies would be faced with difficult business decisions. The worst-case scenario for the town would be if the companies chose to leave Wallingford for another town or state. The town was already reeling from the announcement that one of its largest taxpayers, Bristol Myers Squibb, was relocating to New Jersey. At the same time that the companies identified a need for present and future staff in the STEM fields, our students and families were beginning to discuss the need for increased programming in these areas as well. The challenge the district faced was not vision for such opportunities but the financial resources to bring them to fruition.

The true opportunity that presented itself was that in order for companies to get a well-educated workforce, the school district needed to provide the programming in which to do so. This intersection of the two needs was where the magic happened.

Based on this emerging realization that in order for business and the school district to reach their goals we needed to increase our partnerships, the idea of proclaiming the Town of Wallingford a STEM Town was born. When the idea was presented to the group, consisting of parents, business leaders, community leaders, and educators, many asked, what is a STEM Town, and what is the criteria to be named one? The response was easy to state based on the prior discussions and thoughtful conversations.

The STEM Town designation means that we would all focus on the problem of STEM education and workforce development and will work collaboratively to make a greater impact than if we worked in silos on our own. By coordinating efforts to increase student awareness and interest in STEM fields before, during, and after school, we planned to impact the number of students who pursue such pathways in our high schools, college and the world of work. In addition, by doing so, we believed this increased awareness in students would also impact families. By focusing on the opportunities that students have before them in our town, state, and nation, we hoped parents would support our efforts by supporting their children.

Lastly, we believed local businesses would reap the benefit of a STEM Town designation through the mindset shift that needs to occur in assisting all citizens in their deeper understanding of the many fields and industries in Wallingford. We also hoped that by putting forth publicly that Wallingford is a STEM Town, businesses investigating a move into the town would choose it due to our community commitment to this mission.

The excitement around the STEM Town launch was infectious. Working with students from Quinnipiac University, a documentary was created. The documentary highlighted the needs of the school district and community. It offered a wide range of voices from students, parents, community leaders, business members, and educators. Its premiere became the event of the fall and was an incredible success. The documentary was important because it not only celebrated the hard work and collaboration that was occurring across the systems within the town but also publicly declared we were doing this. There was no going back. Now, all that was left for us to do was to make the impact we intended to have on our community. Little did we know that impact and how it would impact the state.

## STEM Town or Bust

During the ensuing days, weeks, months, and now years after the first showing of the STEM Town Wallingford documentary, success grew. Key areas of success fell into a couple of categories: educational initiatives and community/business initiatives.

*Educational Initiatives*

The district embedded the development of new Career Technical Education coursework within its strategic plan for implementation. Recognizing key areas of need for businesses within the town, focus was placed on Advanced Manufacturing and Medical Careers. The Advanced Manufacturing courses were developed with business members at the table offering the authentic feedback necessary to ensure relevance for the students. The expansion of Medical Careers courses meant an increased number of graduates exiting high school with the nursing certification and the necessary clinical hours for entry into a wide range of medical fields.

In addition to the coursework development and implementation at the high school level, we wanted to make sure that we grew the interest in STEM as early as possible in our system. Working with career and college counselors, the district STEM coordinator, and the district CTE coordinator, a wide range of opportunities for student exposure and exploration were developed. The key was maintaining authenticity in each experience for our students. We did not merely want to create another career day. We wanted students to have an embedded experience of exposure to STEM throughout their entire educational journey in our district.

Messaging all that we were doing in our district and community was essential. A creative way to integrate the schools and businesses was a community-wide STEM scavenger hunt. This took place over the summer, culminating in a town-wide event. The key to this STEM scavenger hunt was that families were encouraged to complete the challenges at local manufacturers at which time they could receive tours and learn more about the businesses. This was very important to all of us in our desire to let our community know how safe manufacturing is and what amazing career opportunities lie ahead for their children. This event was a huge success. Based on this result, design challenges were integrated into the entire district on a monthly basis. STOP, DROP, and DESIGN was another great way to get students at all ages (K–12) to take a moment to engage in fun STEM experiences.

*Community/Business Initiatives*

The community jumped into the STEM Town initiative with great enthusiasm. The mayor made a declaration with representatives from the state legislature presenting a proclamation signed by the governor. The Economic Development Commission started using the title in its messaging to present and prospective companies. Town organizations like the YMCA, Boys and Girls Club, Spanish Community of Wallingford, and the Wallingford Library, developed STEM-focused programming for students and families.

HUBCAP Wallingford, a non-profit community, business, and education partnership, became the epicenter for an employment pipeline program for adults and graduating high school students. Working with local industry leaders, the volunteers at HUBCAP with the Wallingford Public Schools provided six-week training sessions resulting in an opportunity for an interview and employment at the final session. The classes focused on employability skills, resume writing, and interview skills. Candidates did take a math and reading assessment. If necessary, remediation was provided.

## Did it Work?

You may be asking yourself, "Wow, this sounds great, but what has the impact been? Did it really work?"

I am proud to say emphatically, "YES!" STEM Town made an incredible impact on the Wallingford community and continues to do so. Some of the key success indicators came in the places we had planned—education and business.

*Education Successes*

Key educational indicators to success of the STEM Town initiative include the following:

- Implementation of an Advanced Manufacturing

pathway consisting of four courses articulated with Goodwin University

- Increase in Advanced Manufacturing Certificate from the Connecticut Department of Labor to over sixty students annually
- Expansion of medical careers programming to three full-time staff members tripling the number of student candidates for the program
- Establishment of the only United States based International Space Station in collaboration with the Victorian Space Science Education Center in Melbourne, Australia
- Over $175,000 of donations annually for STEM resources
- 2019 Thomas Champions of Industry Award

*Business Successes*

Key business indicators to success of the STEM Town initiative include the following:

- First increase in the grand list in over ten years
- Expansion of international companies in town
- Increased collaboration and support from local businesses
- Employment of 80 percent of the candidates from the HUBCAP Pipeline Program
- State and national recognition for its work around community workforce development

## STEM Town Takeaways and Beyond

I am pleased to report that STEM Town continues to be alive and well in Wallingford, Connecticut. The community, business, and education partnership is strong and mutually rewarding. There is a great sense of pride in the work that was accomplished, but there is also a healthy sense of continued urgency to grow and be ahead of the curve for students and businesses.

There are a few key takeaways from the entire STEM Town experience:

*Dream Big*

The most important thing is to dream big. STEM Towns did not exist before ours. We did not let that stop us. We just used that as a reason to go even bigger than originally imagined.

*Never Underestimate the Power of a Community*

I continue to smile in amazement at the power within the community. Once a common goal was established, everyone was onboard and wanted to help. From donations of money and time to just a friendly call of encouragement, galvanizing people around an important common cause does work.

*Keep Students at the Center*

Always maintain your eye on the prize: student success. In the end, that is all that matters. Listening to students and their needs is essential in getting their buy-in as well as their families.

On a personal note, I have moved on to assume the role as Superintendent of the Goodwin University Magnet School System. In this position, I not only work with schools within

our system but also spend a tremendous amount of time partnering with districts across the state on the development and
implementation of unique STEM programming. The new
position has been particularly rewarding and exciting because
I continue to work with Wallingford as we plan even more
exciting new initiatives around STEM and advanced manufacturing.

# About the Author

DR. SALVATORE MENZO

Dr. Salvatore Menzo, based in Connecticut, is in his twenty-ninth year in public education. Sal has spent more than half this time as a superintendent across three districts. Starting out as an educator in a socially and economically diversified school district, he recognized the need to establish partners for his students with local businesses and resources. This philosophy influenced his work as a Connecticut superintendent in Marlborough for over four years and twelve years in the Wallingford Public Schools. Now, as the superintendent of the Goodwin University Magnet School System, Sal maintains a focus on STEM and workforce development. He continues to apply a collective impact approach, collaborating with local businesses, community organizations, and state-wide groups, to develop a nationally recognized framework for employment pipelines.

In his work in workforce development, Sal has been recognized as an educational leader locally, nationally, and internationally. He has been called upon to speak at many state, national, and international convenings on workforce development as a resource and example for others. He was a featured speaker at the White House regarding workforce development for defense supply chain companies. He was recognized by the

Thomas Industries North American Manufacturing platform as one of the 2019 Champions for Industry. In addition, as the only educator, Sal has been appointed by the Secretary of Commerce to the Connecticut Export Council. He is a member of US Senator Chris Murphy's Manufacturing and Aerospace Advisory Council. Over the last three years, Sal led an international aerospace collaborative with Melbourne, Victoria, Australia. This partnership is providing one-of-a-kind STEM experiences for students while engaging local businesses in the development of pipeline programs for students as potential future employees.

With a strategic mindset situated in education with an understanding and appreciation for business, Sal is committed to working with other companies and communities to assist in ensuring their economic sustainability and growth. He is very passionate about preparing high school graduates for success while supporting the employment needs of industry.

Sal received his undergraduate degree from Connecticut College and the University of Sydney. He was conferred his master's and doctoral degrees from the University of Connecticut.

# THIRTEEN

# Bringing Dreams Closer to Students

## AN OVERVIEW OF AN AMBITIOUS AND REWARDING PLAN

### Dr. Sito Narsissee

East Baton Rouge Parish Public Schools serve more than forty-one thousand children and families in a mid-sized urban district. As the leader of this district, I am constantly thinking about the ways to best serve them to prepare them for the world after they leave our classrooms and how we can bring their dreams closer to them. The East Baton Rouge Parish School System (EBRPSS) has recently embarked on a plan to bring dual enrollment opportunities to every high school student across the district while expanding student's exposure to real-world STEM learning through expanded partnerships with business and industry partners.

## Pathways to Bright Futures

The goal of the District has been to provide more opportunities and access for students and their families. Under my direction, the EBRPSS has embarked on an ambitious and rewarding plan to bring dual enrollment opportunities to all students across the district. Dual enrollment allows students to simultaneously satisfy core coursework and obtain college credit. Through the Pathways to Bright Futures initiative, students will be able to choose from one of five pathways of study beginning in their ninth grade year, ultimately providing each graduating senior with the opportunity to graduate high school with the college credit equivalency of an associate's degree or an industry-based credential (IBC) in a high-demand, high-wage job.

Each pathway has been developed in partnership with the local Chamber of Commerce, and align with those careers that will allow students to enter into a high-wage, high-growth career in their own community. Each student will choose from the following pathways: Technology, Pre-Med & Allied Healthcare, Construction & Manufacturing, Transportation & Logistics, or Liberal Arts and Management. There is a clear delineation in the pathways coursework, essentially allowing students to progress forward through a STEM or non-STEM pathway. Coursework will be designed in such a way that students will begin the divergence in their junior years, providing them the opportunity for exploration and alignment of their own aptitudes and interests.

What we discovered in creating the pathways is that there is an abundance of jobs available in STEM-related fields. Unfortunately, there was and is a persistent gap in a ready, local workforce and talent pool to fulfill those roles, meaning that oftentimes employers are forced to recruit talent from out-of-state and internationally. The Baton Rouge Capitol Region presents another unique facet when considering talent recruitment; the overwhelming majority of Louisiana residents born in the state remain in the state with very little outward migration. This in itself presents a unique opportunity to ensure that the school system is built in such a way that it meets the current and future needs of citizens and employers for years to come. In essence, we have the opportunity to design a school system that is geared to meet the needs of the future workforce, our students, and employers across the capitol region for years to come. By creating a system that provides a direct pathway to employment in high-demand and high-wage jobs *or* providing the opportunity for students to earn multiple, transferable college credits from Louisiana higher education institutions, we ensure that our young adults are able to envision a future for themselves where they can remain successful and contribute to the success of their families for generations to come.

The beauty of the Pathways to Bright Futures Initiative is that it will provide students with the opportunity to obtain college credits in general education courses, providing them with a head start, whether they are college bound or interested, instead, in pursuing a career immediately upon graduation from high school. By scaling the Pathways to Bright Futures work to reach students district-wide, we also work to close the pervasive equity gap that plagues large urban districts and provide more opportunities to expose those students who are under-represented in STEM careers to a pathway to career and prosperity.

## Creating Space for STEM

Baton Rouge, like most urban centers, is a resource rich city. As home to three major hospitals, several industrial, petro-chemical and manufacturing plants, and three major secondary education institutions, the opportunities that we can provide for our students through thoughtful partnership should be accelerated, coordinated, and integrated. Across the city we must strive to innovate and coordinate our STEM partners into the teaching and learning environment.

As we adapt classrooms to meet the needs of the learners of the 21st century, it has become abundantly clear that students must see the relevance in their coursework. Math and science concepts in particular should not be relegated to the world of the abstract. In order for students to understand STEM concepts, they must see firsthand how those concepts can and do function in the real world. One strategy that our school system has employed is to explore how each dual enrollment pathway is purposefully integrated with an external business or organization, capable of partnering with the system to foreshadow real-world experiences.

Let's examine for a moment how this looks in practice for a group of students from East Baton Rouge Parish Public Schools over the summer of 2021. EBRPSS is home to the EBR Career and Technical Education Center (CTEC). Students attend classes at CTEC as part of their supplemental course work and focus on a career pathway, often earning certifications and credentialing in their chosen career paths. Examples of these career paths include IT Networking and Programing, Manufacturing, Instrumentation, and Allied Healthcare fields. During the summer of 2021, five EBRPSS students from across the district, who were pursuing coursework at CTEC, were given the opportunity to participate in paid internships, in their chosen field, with industry partner ExxonMobil. Students put the STEM concepts they learned in the classroom into practice in the real-world and were compensated for their time, talent, and expertise.

The goal now is to explore how it would look for students across the district to be exposed to these types of hands-on learning opportunities, both inside of the school building or on-site in the real world. Our city is one that remains resource rich with a wealth of business and industry partners, as well as wrap-around services, community organizations, and supports.

By leveraging existing resources, forging new partnerships across the city, and creating openings for the private and non-profit sector to connect within the school system, we can and have continued to provide real-world exposure and opportunity for students to lead through hands-on work.

## The Future of STEM

The goal of the Pathways to Bright Futures initiative has always been to bring children's dreams closer to them and to support them in seeing themselves in a career past the four walls of their K–12 education in a way that is concrete and not lead by abstract concepts. By creating STEM pathways that align to tangible jobs in a child's community, students in the EBRPSS are able to envision themselves as the future of STEM.

The field of STEM study in practice is one that is often hands-on and interactive for students, often taking the classroom outside or offering hands-on lessons. By introducing a child to a STEM-related pathway, rigorous and relevant dual enroll-ment coursework, and job exposure opportunities we have made the field of STEM attainable and sustainable while simultaneously empowering a work-ready talent pool.

Access and opportunity should never stand as a barrier for a child. Exposure to STEM curriculum and a potential career in a STEM field can open a world of opportunities for a child. By exposing students early to relevant and appropriate STEM education through dual enrollment opportunities and providing real-world experience for them in high-demand, high-wage STEM careers, we continue to prepare the work-force of tomorrow, equip our students with the tools they need now for the STEM careers of tomorrow, and ultimately bring children's dreams closer to them.

# About the Author

## SITO NARCISSE, ED.D.

Dr. Sito Narcisse serves as the Superintendent of Schools of the East Baton Rouge Parish School System. Dr. Narcisse most recently worked as the Chief of Secondary Schools of the District of Columbia Public Schools. He understands the challenge of being a young student trying to learn English and living between two cultures, all the while adapting to the US public education system. The son of Haitian immigrants, Dr. Narcisse moved with his family to Long Island, New York, in the pursuit of a better life for him and his siblings. As an English language learner, Dr. Narcisse learned to navigate both the social and academic obstacles that confront millions of students today. His success as a student led him to enroll at Kennesaw State University in Georgia. Seeing his second language as a strength, Dr. Narcisse graduated with a degree in French and pursued a master's degree from Vanderbilt University in secondary education. Doctoral studies led him to the University of Pittsburgh, where he earned a doctorate in educational administration and policy studies and leadership.

Serving as both a teacher and a principal, Dr. Narcisse opened a high school in the Pittsburgh Public Schools and led turnaround efforts in a Boston public high school. Dr. Narcisse has also been a director of School Performance and acting Chief

School Improvement Officer for Montgomery County Public Schools in Maryland, an Associate Superintendent overseeing school improvement efforts for 74 schools in Prince George's County Public Schools in Maryland, and the second officer in charge as Chief of Schools for Metro Nashville Public Schools with 159 schools.

FOURTEEN

# The Importance of STEM to Post-pandemic Pedagogy and Community Building

Dr. Meisha Porter, Ed.D.

In March of 2019, when the COVID pandemic forced the New York City Department of Education to shut the doors to in-person learning, I was the Executive Superintendent of the Bronx. The Bronx serves students who are 83 percent Black or Hispanic and has been plagued for generations by persistent racial disproportionalities. Many diverse cultures and ethnic groups have called the Bronx home over the years, creating culture-rich communities. Bronx schools serve 190k+ students, 60 percent of our students are Hispanic, 27 percent are Black, 5 percent are Asian, and 6 percent are white, and 80 percent of our families live in poverty. When in-person learning stopped, the enormity of the disparities in our communities became clearer through the digital divide. Schools who were ready sent their students home with devices only to uncover that many lived in communities where Wi-Fi access was limited, unreliable, and, for students in temporary housing and/or shelters, simply not present. Simultaneously, a new definition of an "essential worker" emerged. Many of the parents and families of black and brown students were relegated to a category that placed them at greater risk of being exposed to COVID-19. Spikes in cases of anxiety, depression, and suicidal ideations ensued as students worried about the impact of COVID-19 on their working parents. Parents had to choose between staying home to support their children as they navigated this new normal of remote learning or working to provide for their families. Most chose the latter to ward off the increased level of food insecurity that continues to plague communities where students rely on school for meals.

In March of 2021, I became the Chancellor of the New York City Department of Education. I became responsible for the largest school system in the country and ushering in a new school year. As I took on this enormous task, I had three priorities that were really one: Open, Open, Open. First, open high schools for in-person learning as we had not yet done that in the 2019–2020 school year. Second, open a summer program like none other that would integrate academics, enrichment, and social emotional supports. Finally, and most importantly, open the largest school system in the country for in-person learning for all students by September. One of my major learnings from the pandemic was a true understanding of what the 21st-century classroom should look like. If you are in my age group and were in school in the 1980s, it was about the technology in the "computer lab."

*1980s computer lab*

Over the decades, this moved from Mac labs to smart boards to laptop carts to iPads. However, the lesson from the pandemic was simple but complex, a true 21st-century classroom is leveraging modern technology to enhance the interactions between students and teachers across all disciplines. The creation of this 21st-century classroom allows students to explore content while developing authentic, media-rich projects that deepen knowledge and allows students to connect to the world beyond the classroom. If we develop these classrooms, we will prepare students for the future of our tech-driven world and ensure students are engaged and excited about learning. The key to this: STEM!

Often when I discussed the reopening of the largest school system in the country, it was through the frame of reference that we could NOT go back to normal. The normal that we knew perpetuated the inequities that were heightened by the pandemic; STEM presents the opportunity to a pathway away from that normal. STEM education is the intentional integration of science, technology, engineering, and mathematics, and their associated practices, to create a student-centered learning environment. An environment in which students investigate and engineer solutions to problems and construct evidence-based explanations of real-world phenomena with a focus on a student's social, emotional, physical, and academic needs through shared contributions of schools, families, and community partners. STEM is also a methodology that encourages students to solve problems that are relevant to the world in which they live. STEM classrooms flip students' learning experiences to mimic real-world scenarios and engage students in problem-solving in a way that allows students to aspire to solve problems in their communities. This point was driven home during a visit to the New York Harbor School on Governors Island, where students gave me a tour of their Billion Oyster Project aimed at restoring one billion live oysters to New York Harbor! Not only was this hands-on work, but it also required discussions of ecosystem restoration and theoretical concepts that are best done in person. The natural curiosity of students is nurtured when they are investigating scientific concepts and phenomena using simulations and virtual labs. I also got to paddle and learn with students in the Bronx on the water at Roberto Clemente State Park in celebration of the tenth anniversary of the Urban Waters Federal Partnership on our amazing community waterways. Through these tours, students got the chance to learn about the geography and history of river while paddling in it. Afterward, students sat in the park and engaged in water quality testing, seed planting, and learning to tend to the indigenous plants in their park in their community. This is an example of transitioning student learning to project-based learning that truly connects students to real-life issues affecting them and their communities.

*Paddling along the Harlem River with Bronx students*

In the STEM framework published by the New York City Department of Education in 2015, they reference an editorial by the New York Daily News, which notes that "by 2020, the US economy will demand 123 million high-skilled workers with strong backgrounds in science, technology, engineering and math." The problem is, according to the same article, only 50 million Americans will qualify for those jobs. The larger question is, who will those Americans be, what communities will they come from, and who will they represent?

As I exit the New York City Department of Education and the role of Chancellor, I am excited to be returning to my home borough, the Bronx, where I will take on the role as the inaugural President and CEO of the Bronx Community Foundation. The Foundation is focused on building an equitable, inclusive, and just Bronx. A key priority of the Foundation is to improve access to modern digital resources for Bronxites. For many years, the Bronx has been in the "digital desert"; ensuring strong technological connectivity has the potential to disrupt the longstanding inequities that exist in the borough. The ability to ensure residents and families have access to online education and training, employment, social and civic engagement, financial resources, and health have been deeply tied to the internet access in the 21st century.

Households in the Bronx have a median annual income of $41,432, which is less than the median annual income of $65,712 across the entire United States and $68,304 in New York City. In 2019, median earnings for full-time, year-round workers ages twenty-five and older in a STEM job were about $77,400. A report from the Organization for Economic Cooperation and Development estimates that the global closure of schools could lead to a 3 percent lower income for K–12 students over their lifetime and a corresponding average of 1.5

percent lower annual gross domestic product for countries for the remainder of the century (Hanushek and Woessmann 2020).

At Almost

# 38%

THE BRONX IS THE BOROUGH WITH THE HIGHEST
PERCENTAGE OF RESIDENTS WITHOUT HOME BROADBAND

The Bronx has the lowest broadband adoption rates of any borough and the disparities are even more pronounced at the neighborhood level.

NEARLY 1-IN-5 TEENS can't always finish their homework because of the digital divide.

Roughly one-third of households with children ages 6–17 and whose annual income falls below $30,000 a year do not have a high-speed internet connection at home.

COMPARED WITH

**Just** of such households earning $75,000 or more a year.

BASED ON THE MOST RECENT FCC DATA, LARGE SECTIONS OF THE BRONX DO NOT HAVE A 1,000 MBPS OPTION. THIS SAME AREAS IS ALSO LIMITED IN THE NUMBER OF SERVICE PROVIDERS FROM WHICH TO CHOOSE.

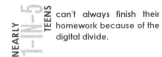

The New York City Internet Master Plan | Pew research center

A report from McKinsey Insights estimates that the average K–12 student in the United States could lose the equivalent of a year of full-time work income over the course of his or her lifetime, and these losses may be higher for Black and Hispanic students (Dorn et al. 2020). Competitive wage careers increasingly depend on a workforce with advanced science, technology, engineering, and technology skills. We also know that STEM employment in the United States continues to grow at a faster pace than employment in other occupations, STEM workers command higher wages than their non-STEM counterparts, and STEM degree holders enjoy higher earnings, regardless of whether they work in STEM or non-STEM occupations.

To shift this narrative for Bronx students and families, a comprehensive approach to STEM education along with the Foundation's commitment to focusing on digital equity across the borough is necessary. Addressing the glaring inequities in the achievement gap in STEM education is critical. According to the National Science Board, more than half of Asian and white students across all grade levels score at or above average on STEM standardized tests, compared with only 28 percent of Hispanic and 18 percent of Black students. In addition, research also tells us that for many low-income students, science knowledge gaps exist even *before* kindergarten—and widen dramatically in primary and secondary school. STEM subjects need to be a part of core instruction at the elementary level, so students have the Foundation to succeed in STEM subjects in middle and high school. As we consider our population and demographics, it is critical that we center student learning in acknowledging our students' and families' lived experiences through STEM. This too aligns with the Foundation's commitment to invest in the power of the community to eradicate inequity and build sustainable futures for Bronxites with Bronxites.

So, what must we do? I am sure I don't have to tell you that STEM is critical to the future of our city and nation. I am also sure I don't have to tell you that in-person collaboration and discourse are what deepen students' understanding of STEM concepts and practices. STEM cannot be an add-on; it is essential to prepare students for college and careers. As Chancellor, I was able to leverage resources through the Academic Recovery Plan to guarantee every K–12 student has access to an LTE-enabled digital device with Wi-Fi. We also committed to expanding access to Computer Science 4 All to four hundred thousand students by 2024, training over five thousand educators in advanced computer science and launching a technology capstone project for all eighth-grade students to demonstrate digital literacy. School systems must dedicate more time and resources to STEM at the elementary level. We need to provide ALL students with the tools and knowledge to be successful in STEM so that they are ready to continue their STEM education in college or to enter STEM careers after high school. Through my work at the Foundation, I look forward to ensuring Bronxites have access to broadband and digital resources by eradicating the Wi-Fi deserts that exist in the Bronx and similar communities. Another bit of real talk: STEM fields have long been gatekeepers in education, serving to limit the number of people who have access to higher education. So, it is our duty to work together to reimagine this field as a gateway to interrupting poverty by being intentional around providing access to students from underserved communities to a path toward lucrative STEM careers. And let's remember that all young children are naturally curious about their world. So, it is our responsibility as education leaders to develop and promote research-based practices that leverage this natural curiosity.

# References

Dorn, Emma, Bryan Hancock, Jimmy Sarakatsannis, and Ellen Viruleg. 2020. *COVID-19 and Student Learning in the United States: The Hurt Could Last a Lifetime.* June 1. https://www.mckinsey.com/industries/education/our-insights/covid-19-and-student-learning-in-the-united-states-the-hurt-could-last-a-lifetime.

Hanushek, Eric A., and Ludger Woessmann. 2020. *The Economic Impact of Learning Losses.* September. http://hanushek.stanford.edu/sites/default/files/publications/The%20Economic%20Impacts%20of%20Learning%20Losses_final_v1.pdf.

# About the Author

DR. MEISHA PORTER, ED.D.

Dr. Meisha Porter will be joining the Bronx Community Foundation as the inaugural CEO/President. Meisha recently served as the Chancellor of the New York City Department of Education, the largest school system in the nation. She is responsible for educating 1.1 million students in over 1,800 schools. Prior to taking on the role as Chancellor, she served as the Bronx Executive Superintendent where she was deeply invested in sharpening school leaders' equity lens and building collaborative practices across schools. She also served as the Superintendent for Community School District #11 and Principal of The Bronx School for Law, Government and Justice (LGJ). During her tenure at LGJ, Meisha served as Community Coordinator, Internship Coordinator, and taught English before becoming an Assistant Principal and then taking the helm as Principal in 2004. Having worked her way up through the ranks, she is exceedingly aware of the challenges city schools and communities face and has dedicated her life to improving the learning environment for all students. Meisha also believes that if school leaders and teachers work collaboratively with parents and community partners to focus on empowering students to achieve their unique individual potential, students will be prepared to meet the demands of the 21st century.

Born in Far Rockaway and raised in Jamaica, Queens, and the South Bronx, Meisha began her life just as her students do. A product of the New York City Public Schools, she graduated from Queens Vocational and Technical High School and went on to receive her Bachelor of Arts in English concentrating in Cross Cultural Literature and Black and Puerto Rican Studies at Hunter College. She completed her School District Leader certification through the NYC Advanced Leadership Institute, received her master's degree in Administration and Supervision as well as an honorary doctorate in May 2020 from Mercy College and most recently completed her EdD at Fordham University. Meisha lives in the Bronx with her husband Norris Porter, her daughter, Jordyn, and honored daughters, Deshauna, Lakara, and Jydin.

# FIFTEEN

# Leaders for Learning

## MANAGING AMBIGUITY IN THE PURSUIT OF EQUITY IN OPPORTUNITIES AND OUTCOMES

### Dr. L. Oliver Robinson

Leadership is about managing ambiguity. Consequently, it is paramount to avoid the pitfall of binary thinking when facilitating intentional interruption to foster agency, engage in productive struggle, and ultimately broaden the cognitive horizon of those whose lives we impact. The transforming and transcending power of education serves to provide unbridled opportunities in which the credibility of our students as learners is never questioned. Consequently the mission, vision, goals, and even lesson plans of teachers should include a dream plan for every child—a declaration of the commitment of leaders for learning to position each student for prosperity.

It is critical that leaders pause and draw back the blinds of their life and peek through the windows of their private and personal predicaments...keeping in mind some simple questions: Am I doing enough? Have I exhausted the breadth and depth of my talents and abilities to be an asset for someone else, namely the students I serve? Or am I *blinded* by biases, subsequently being a hindrance to someone's pursuit of excellence?

Students are all dreamers, and it is the responsibility of leaders to instill in them the notion to always dream big dreams and strive for extraordinary successes. Students must be reminded every day that as long as their dreams are bigger than their circumstances, they cannot fail, and to fail does not make one a failure. That is why leaders for learning, those who chose to serve children and foster communities of learning, must be stalwart in the commitment to have a resounding and positive impact.

As leaders for learning, the role is to help students expand their circumstances to fit their big dreams, to be difference makers. The point should be punctuated that leaders for learning must live each day with a strident sense of purpose, knowing that the realization of the dreams of each and every student is indeed an intentional act. Such a commitment of conviction is particularly vital now in an environment marked by constant technological advancements and increased global-ization. Increasingly, leaders are functioning in uncharted waters, facing unprecedented challenges, and relied upon to acutely manage ambiguity.

## Pivoting the Fabric of Traditional Education

March 20, 2020 was a critical turning point for schools, given the closing of school buildings and the abrupt shift to online learning. The COVID-19 pandemic pervasively altered the very fabric of the provision of traditional educa-tion. School districts were compelled to develop and continu-ously expand upon comprehensive plans for continuity of learning; eLearning Plans became a household name.

Schools were compelled to go from onsite learning and teaching experiences to totally remote eLearning. With that came:

- huge learning curves;
- huge deployment of technology to students and staff; and
- huge development of online content and shifts in delivery strategies.

At the core of such endeavors, students receive instruction based on continuity of learning, with an unwavering focus on providing equity in opportunities and outcomes for all students. The execution of any such plan must be contingent upon ubiquitous access to technology capacity, encompassing clear procedures to ensure reliability and accessibility of technology for students and school personnel alike. In other words, assumptions about availability and access to technology cannot be the basis for decision making. The stakes are too high for students. Ubiquitous access must be verified, and verified for each child in a household. The technological divide is the new institutional discriminatory gap, disenfranchising many students in a technology desert. Consequently, any effort focused on continuity of learning must be prefaced on a model of service following needs to preserve equity in opportunities and outcomes.

The veil of false parity has been lifted, revealing the gross inequities that exist in society and that are manifested in schools. Economic and financial disparities correlate directly with the lack of ubiquitous internet connectivity and technology access. Coupled with lack of parental capacity to effectively navigate the varied and seemingly changing technology platforms, debilitating discriminatory practices are fueled, and resource incapacity is the inevitable outcome.

The pivot to virtual learning keenly illustrated the need for leaders for learning, regardless of title, to diligently work to contend with one of societies' greatest dilemmas: the closing of the academic achievement gap between black and white, rich and poor...preserving equity in opportunities and outcomes, parity in performance, and high quality learning experiences for *all* students.

Educational institutions are in the midst of pivoting and changing the revealingly archaic model of seat time and lock-step progression of achievement marked by assessment bench-marks. The pivoting of education in this era is both tactical and strategic...keenly analyzing and addressing the impact of technology on positioning school systems for continued prosperity.

## Progressive Leadership Is Not Binary

Continuity of learning remains the North Star, the true north, the fixed point in a seemingly spinning equation for societal normalcy. Education remains oriented to the fundamental value and belief of preserving equity in opportunities and outcomes for all students. Correspondingly, the vast differences in student needs from special education to acceleration, the provision of support services from ENL to mental health and wellness are illuminated. Even the gap in educational achievement itself is firmly appreciated as being merely a symptom of a larger problem. It is, in many ways, a manifestation of gross disparities in society, the opportunity gap that stems from discriminatory practices and the willful marginalization of certain individuals and communities. Some also call it opportunity debt...recognizing that there is a sense that not only was opportunity denied, but it was denied despite being owed. Far too many children are victims of the unfortunate circumstances of their birth, living on the peripheries, at the outer edge of disappointment and despair...disenfranchised.

In these crucible times, where lives are being transformed daily, the fundamental values and beliefs that actions must always be focused on the best interest of students must not only drive the provision of education but fundamentally reform it. There is no playbook to help contend with the challenges being encountered, only a steadfast commitment of conviction to do what is right and preserve equity in opportunities and outcomes for all students.

The complex nature of learning and the critical need for differentiated approaches to capture and engage students speaks to the need for demonstrative changes. The propaganda of academia can no longer be the running narrative. The old model of offering to serve but not effectively serving all students must be called into question. The seemingly innate connection between high performers and low achievers must be shattered, for the endorsement of one is to reverse the acceptance of the other. The integration of technology allows for the prioritization of learning standards, a slowing down of the proverbial game and allowance for deeper, independent engagement with content and multiple measures of proficiency. It is about meeting students where they are and taking them to where they ought to be, endeavoring to remove all the contradictions between those two points on the learning continuum...where they are and where they ought to be.

As we lean forward, viewing the conditions and provision of schools prospectively, assessment of needs must be done through an equity lens with an undeviating eye on the WHY, the provision of Authentic Learning, based on tenets of education being Inclusive, Diverse, and Axiomatic:

- Inclusive: What else can we do to make the school and classroom experience more relevant?
- Diverse: How can we be more responsive to and capitalize on emerging changes and cultural shifts?
- Axiomatic: What else do we need to do to ensure equity in outcomes, *all* students succeeding and realizing their full potential?

A keen focus must be on organizational resiliency...finding the flow, the zone of seamless performance and productivity. Doing so necessitates leadership fortitude, which plays out in four dimensions:

1. Contending with systemic change—process and practices revamped to be rigorous and relevant, or even trashed and replaced
2. Mindset shifts—not succumbing to being prisoners of past assumptions, fostering resiliency to embrace the challenge of change
3. Personnel retooling and recalibration—practices becoming the common practice, ground in a sense of responsibility versus responding to external measures of accountability
4. Contingency and scenario planning—allowing for seamless organizational pivoting in a fluid and uncertain environment without losing momentum

As leaders for learning, we cannot succumb to the sucker's choice of a binary view of progressive leadership. Technology is the bridge to knowledge and, consequently, a vital component of the eradication of gaping performance disparities. Access to and knowledge of effective use of technology can exponentially move the proverbial needle of progress toward parity in outcomes. In a knowledge economy, fueled by technological advancements, a comprehensive STEM approach to education instills in our students a sense of significance and cultivates the ability to be self-directed learners, to be innovative thinkers.

## Doubting Objectivity: The Horizon Leans Forward

The crux of the matter is that people have a natural inclination to lean toward avoiding discomfort, hence defensiveness, or even engaging in dissonance...believing one thing, yet doing another. This is hard work, complex work. Why? One of the most complicated components of being honest in our deliberations, diligent in our assessment, and acting with fidelity is the need to consistently doubt objectivity, meaning confronting the truths of current circumstances. It means challenging the subconscious that results from social conditioning, lived experiences. Doubting objectivity means being audacious enough to ask critical questions and to scrutinize ideas, constructs, traditions, even our seemingly innate beliefs.

Here is the end game. Doubting objectivity is about dismantling, disbanding discriminatory institutional practices that lead to the perpetuation of uneven outcomes and hinder the highest level of achievement for all students. It must all be a pursuit of Alpha, a positive return on the investment. That means sanctioning the inevitable. Being arrested by a commitment of conviction, possessing a keen sense of intentionality to challenge and change the status quo, which in itself is a luxury that paralyzes progress. It is critical to confront the reality that, in many ways, schools are merely creating test-taking soldiers and minimally compliant learners. Doing so creates seismic shifts toward STEM education and the cultivation of innovative thinkers. This paradigm shift manifests in cultural conditioning that defines expectations, norms for actions, impetus for evasive measures to pierce the status quo, the natural enemy of progress, and consequently redefines new plateaus of progress.

The horizon for our students leans forward, and the gates of a new era have been flung wide open. Leaders must be directed by the optimism of the will and not by the pessimism of the intellect. So, when you draw back the blinds of your life and peek through the windows of your private predicaments, ask yourself, Am I doing enough, or am I blinded by my own biases? What's your answer?

# About the Author

## DR. L. OLIVER ROBINSON

Dr. L. Oliver Robinson is the author of *Naked in the Public Eye: Leading and Learning in an Era of Accountability.* His recognition as the 2013 New York Superintendent of the Year speaks to his keen leadership. Dr. Robinson is one of the longest serving superintendent of schools in the country. He has served as the Superintendent of Schools for the Shenendehowa Central School District since July of 2005. He also served as the Superintendent for the Mohonasen CSD from 2001–2005. Dr. Robinson also serves as an adjunct professor at the Sage Colleges and the College of St. Rose. He received his bachelor of arts from Brown University, and a master's and doctor of philosophy degree (Ph.D.) from the University at Albany.

SIXTEEN

# Light the Fire

Mika Salas

Conduct a short social experiment for me. Ask twenty people what made them choose their current career. The twenty people can be from any demographic, they can come from different backgrounds, and they can hold any position in their organization. Ask them why they chose that type of work and who or what inspired them to pursue it. If you are anything like me, you will be surprised by their answers.

I grew up in a small rural coal mining community. They say for every coal miner there are nearly ten supporting jobs in town. My family of three was supported by one of those other jobs as my mom was a nurse. She was my only parent, and she worked very hard to provide for my sister and me. Unfortunately, her pay did not match her determination and grit.

Growing up poor did a lot for me. It provided a lifetime of motivation and a deep passion to do whatever I could to successfully bridge the chasm from poverty to middle class. It has also driven me for nearly twenty-five years in education to help our students do the same if they choose. Sharing the details of overcoming the effects of poverty is for another book, but we as educators must realize that all students are seeking success, and they come to us with their own unique set of talents and interests. Our job is to provide the opportunities and experiences that may ignite a passion deep inside of them.

I remember being in junior high and my teacher chose about ten of us to stand in the front of the room. We were asked to arrange ourselves from shortest to tallest. It only took a few seconds, and there we stood neatly arranged as she requested. She asked the class to identify the leader of the group, and I thought it was a very silly question. There was no leader. We almost magically arranged ourselves in only a few seconds like a work of art. I was genuinely shocked when everyone pointed at me. I didn't realize it, but I had arranged and executed the plan to accomplish the task. It was the first time anyone called me a leader, and it changed the way I looked at myself. In my head, I was no leader. I was the kid in off-brand clothes who laid in bed every morning listening carefully and hoping that my mom's Pinto would start so she could make it to work on time. We did not have what others had, and I thought it was impossible that other students could look at me as a leader. I needed their perspective to begin seeing myself as something more than the poor kid with no options. It was one small activity in one small class, but I am forty-five years old and distinctly remember how proud I was that my classmates pointed at me that day.

Poverty is not the only hurdle. There is a wide range of issues facing students. When you listen to teenagers, you understand that they are seeking direction and options for their future. Many teenagers have difficulty sifting through the endless amount of information out there. Sometimes having so much to choose from leads to indecision and frustration, then no choice at all. This is particularly true when students are left on their own without intentional conversations with adult mentors or hands-on exposure to new things. They do not yet know what industry excites them, what type of work they want to do, or what lifestyle they would like to lead. They only know what they have been exposed to and might not realize what can be made available to them.

Our role as educators is to provide experiences beginning in elementary school that guide students through a variety of career options to help them develop and hone their natural abilities while inspiring them by igniting a passion for their future work.

An effective model for quality STEM education includes vertical alignment beginning in elementary school and continuing through junior high, high school, post-secondary, and into employment. In elementary school, students can be tasked with creating, building, coding, and solving open-ended problems regularly. They can be visited by guest speakers representing a variety of industries. These presenters may be asked to describe the type of work they do, details about their industry, and the education required for a range of positions in their respective companies. They could bring examples of their product and give the students a way to interact with them and the work they do. It is very helpful for students to consider their own personal use of the product and how it makes their life better. For many students it will be the first time they consider jobs beyond what their parents do for a living. They will start to look at their community differently and begin to wonder what types of jobs are out there. The expectations they have for themselves will begin to grow, and their world will begin to expand.

It is also important that we praise students for all aspects of what they offer the world. We need to find and develop attributes that make great employees and citizens and not focus solely on the academic success. There are future scientists, computer programmers, engineers, and entrepreneurs in every classroom, but they may face barriers because of some aspect of their lives, and we must break those barriers down for them.

In junior high we build from the elementary experiences. Junior high students benefit from field trips where they can see, smell, and touch the different businesses. We can use guidance counseling software to arrange field trips that align with natural curiosities, interests, and strengths. Students can begin to see that there are a wide range of career options for them within each industry. For example, they may think construction is the most appealing industry. After visiting an engineering firm, they may discover a whole new world of planning and design that they never knew existed. A student will have met an engineer who can later arrange for an internship for them in high school. These experiences coupled with an exploratory career course and elective options give students a variety of avenues to consider. There is something special about physically walking around a business and talking to employees that is much more effective than limiting exposure to an internet search or class discussion.

As students prepare to register for high school, they should take a tour of the school. They should be expected to physically visit each department and interact with teachers and students. They can see examples of artifacts or products created in each department. The goal is to avoid course selection based on friends' schedules or teacher shopping. We should also begin the conversation about internships, apprenticeships, certification, and employability for skills mastered in

STEM courses. Students want to know what it takes to obtain STEM jobs and the expected salary for those who choose to pursue them. Our counselors play an invaluable role in helping students align the advanced math, science, and coding courses necessary to pursue these fields. It is much more feasible now with Concurrent Enrollment and online learning options.

Businesses have a vested interest in finding and growing local talent. They are eager to help schools if they are given an opportunity to do so. School leadership should work closely with local business owners, higher education, and technical training facilities to offer stackable credentialing, internships, and apprenticeship programs. Such models allow students to graduate from high school fully certified in a variety of fields. This exposure often creates a change in perspective for students. They begin to expect more from themselves, and they set challenging goals that they might not have considered without the experiences provided through a quality STEM education.

How does all of this connect to the social experiment I proposed at the beginning of this chapter? I believe you will find that nearly all of the people you interview will have landed in their respective career field by happenstance. It may have been a childhood experience that connected them to a specific field or a simple conversation with a teacher who suggested they pursue a path after having demonstrated a skill. The truth is that many of our most important decisions related to career options were made because of a conversation or networking that we did not find on our own. Decisions were made because someone made a comment or suggestion. It was because an adult believed in us and offered a different possibility.

It is critical that we expose students to a wide range of challenges and experiences that allow those types of interactions to take place. We need to treat each conversation as THE time we will change the trajectory for a student because that very well may be the case. We never know what will ignite the fire, but we can strike as many matches as possible.

# About the Author

MIKA SALAS

Taking an opportunity to pay it forward, Mika Salas led a high-poverty middle school from one of the lowest performing to one of the highest in four years as principal. She has worked her way from "one of those kids" to Superintendent of Schools in her twenty-four-year education career. She is passionate about setting high standards for students while adjusting the entire education structure to support them as they reach their goals and beyond. She is a steward of the PLC process, RTI, and finding ways to help students overcome the effects of living in poverty. Her writing is focused on the need to provide students with employable STEM skills that will allow them to span the chasm between poverty endless potential on the other side.

She champions for students and has been recognized as a 2017 Solution Tree Model PLC School, the 2017 Utah Rural Schools Administrator of the Year, and the 2016 Carbon School District Administrator of the Year. She also owns an education technology company, Tiered System Support, which offers a solution for managing flex time in secondary schools.

Website: www.tieredtracker.com

# SEVENTEEN

# STEM Education and CTE Pathways

## CREATING OPPORTUNITIES FOR GENERATIONAL CHANGE AT A LARGE URBAN HIGH SCHOOL

### Rene Sanchez

Cesar E. Chavez High School is located in the southeast corner of the Houston Independent School District (HISD). The student population exceeded three thousand students, with more than 90 percent free and reduced lunch and over 99 percent students from historically marginalized populations. Thanks to the HISD's selection for a US Department of Education Race to the Top (RTTT) grant, the district and the campus were able to re-introduce Science, Technology, Engineering, and Mathematics (STEM) education to the school. The RTTT grant allowed Chavez to provide students education focused on STEM outcomes. To support this, the school created an academic house system based on CTE pathway strands. The purpose of this was for students to develop strong relationships with their instructors and for administrators to monitor their educational and socioemotional outcomes. As the STEM program matured, Chavez added the International Baccalaureate Career Program (IBCP) to serve the students who struggled to choose between the advanced academics of the IB Diploma Program (IBDP) and the Career and Technology (CTE) STEM certificate and licensing programs.

My connection to STEM education is that I served as the Chavez High School principal during these changes to the school. The importance of this chapter shows that a large urban high school can successfully change course and then sustain STEM education. This success is replicable because the district and the community provided enough financial and postsecondary resources and recruited, retained, certified, and supported the instructors needed to staff the programs.

Chavez High School is set in a unique place in Houston. In addition to being close to the Texas Medical Center, NASA, the Port of Houston, several colleges and universities, it sits across a tributary of Sims bayou from a contiguous set of five multinational petrochemical companies, one of which had been a partner supporting the school and its STEM program since its inception in 2000. Near the beginning of my tenure as the Chavez High School principal, I met with a representative from Texas Petrochemical Company (TPC). They shared that they were very interested in having more Chavez students work for their company. She shared that many of its personnel were approaching retirement age, and TPC wanted to recruit Chavez students to provide for its neighborhood and help guide the company's future. We talked about the different CTE offerings in the school and how they could be helpful to this request. One thought that she shared made perfect sense to me. In every district adjacent to this southeast corner of HISD, the neighboring district had an associate degree dual enrollment program with San Jacinto College in Process Technology. Process Technology is a hybrid degree where students learn the technology, operations, and leadership and eventually lead within the plant environment. The representative wanted to be sure that Chavez students had equal opportunity to be selected for TPC's high-paying jobs, just like the students from the neighboring districts.

## Initial Changes

With the backing of the RTTT grant and Houston ISD's CTE department, Chavez High School decided to begin wall-to-wall CTE programming at the school. Chavez's architecture, built at the end of the 1990s and opened in 2000, was designed with smaller learning communities in mind. So, each pathway could have its own area of the building, replete with an administrative suite and teacher workrooms. The placement of the pathways would not be difficult, but the transition to wall-to-wall CTE would prove to be challenging to get started.

At the time, the school had the vestiges of an engineering program that was not producing enough students with certifications or licenses. Also, the program was not sending graduates on to colleges to earn associate or bachelor's degrees in engineering. Students frequently took the courses within Chavez's engineering strand out of order, and the engineering curriculum was not aligned to any end goal. Moreover, neither the school nor the district had updated Chavez's facilities and instructional resources recently. So, to meet the request of our TPC partner to produce more students interested or qualified to work at its or other companies' neighboring plants, the school would have to change how it prepared our students for opportunities after high school.

Early on, the first correction we made in the engineering program was adopting Project Lead the Way (PLTW) to coordinate our curriculum, the training, and the certification of our teachers. The curriculum change took about a year and was not the difficult part of the transition. Rather, once the school addressed the engineering staff, it helped the growth of the program almost immediately. Between retirements and reassignments, we provided a coherent sequence of courses for students to progress through the PLTW pathway.

Chavez had three other STEM pathways in the building. The other pathways include Health Sciences, Environmental Science, and Digital Media. Incoming ninth-grade students became part of the ninth-grade house. In that house, we placed the emphasis on matriculation to tenth grade and the learning of the CTE STEM offerings while being on campus. The Health Science program also began using the PLTW curriculum and coursework. It began producing students who earned their intravenous and Pharmacy Technician certificates. (Given the COVID-19 pandemic, the students who achieved these certificates are, no doubt, high in demand.) The environmental science program included our veterinary and floral technology program. Finally, the digital media program also used the PLTW curriculum and coursework, and it offered opportunities for students to use video editing and graphic design software to create.

One of the many reasons the school experienced this success is that we had wall-to-wall CTE pathways. Every student was assigned a core group of teachers, grouped with a CTE or ninth-grade pathway. These core teachers would have a common pathway planning period to align instructions and interventions. Mind you, to create a master schedule for a three thousand student school with these parameters was a herculean task in and of itself. Assembling the courses, the teachers, the planning and conference periods, the classrooms, and the many other permutations, took a planning team and a great programmer to get it workable. Had we only had a few pathways, not every student would have direct access to a post-secondary option, and not every teacher would have a role in providing career-relevant education to the students.

Concurrently, Chavez was also undertaking the process for IB DP authorization. We wanted to ensure that the students of southwest Houston had the same opportunities and access to advanced academics and the benefits thereof that students from the more affluent parts of the city did. Therefore, once we became an IB World School offering the Diploma Program, we began authorization for the Career Program. This way, our students who had difficulty choosing between STEM career pathway options and advanced academics could have the opportunity to do both. Additionally, as these Chavez students applied for jobs, college admission, and scholarships, they would have the dual benefit of the STEM and IB background on their résumé.

An initial challenge for the new CTE teaching staff was the expectations of the older students to whom the change to career-based houses made scheduling or expectations difficult. As we reorganized students into their CTE STEM houses, students in eleventh and twelfth grade had difficulty adjusting to the new courses and expectations. Many of the students wanted to drop the classes, and several did. Still, the reorganization of the school allowed for a coherence of learning that benefited the students who were in the younger grades and who had yet to enroll on the campus.

## Outcomes of the Changes

Over the next eight years, we transformed a once nascent STEM CTE program into a consistently strong program.

Within a graduating class of six hundred students, the CTE program easily produced over one hundred students who earned Pharmacy Tech, Certified Clinical Medical Assistant (CCMA), Veterinary Tech, Solidworks, Emergency Medical Tech, Fire Technology, and IBCP certificates. In 2019, Chavez graduated its first set of Process Technology associate degree students from San Jacinto College. In 2021, it graduated its first IB CP students. Chavez High School's successes can be found in the belief in the students and the school by the business community, the consistency and inclusiveness in campus leadership, and access to financial, academic, and curricular resources. Without these lessons intersecting at the same time at Chavez, this urban high school would not have provided the same number of STEM opportunities for its students. A school or a district must have a shared vision for outcomes that it expects its graduates to attain. Without one, some students will fall through the cracks of the school or district and not achieve expectations that they, their families, the school, or the community hope for them.

Planning for STEM in your school or district requires several items to consider and assist in long-range success. First, you will need a North Star to guide your way. Setting a vision and creating the outcomes that the students, school, and community desire is imperative. The shared vision will enable the setting of interim goals to ensure progress toward the long term. It will also allow for short-term wins to celebrate and acknowledge excellence. Finally, it will give a common language to all stakeholders to perpetuate the work necessary to accomplish the outcomes from the shared vision.

Second, account for every student in the building, especially those from historically marginalized populations. If they are not involved in a CTE STEM pathway, they need to be involved in something the school offers. That *something* within the school will assist them with engagement and belonging and will ensure that they attend school regularly and graduate on time.

Third, as the shared vision is set, necessary resources must be provided for the teachers, the administrators, and the students. Whether the resources include professional learning, extra duty pay, additional software, equipment, internships, transportations, competitions, or anything else needed to accomplish the shared vision, administrators must make sure that the teachers and students have access to it.

Lastly, once the shared vision is created and all along the implementation process, a school or a district will need to find and address the barriers that reveal themselves. The barriers could come in the form of people, systems, architecture, technology, beliefs, curriculum, or geography. The barriers need to be removed or reworked to align the work of the school or the district to the shared vision. They can impede or completely derail the progress of the critical STEM work.

The administrators and the staff could not have accomplished Chavez High School's success story without using technology as a tool to help us achieve our shared vision. Our school was fortunate to be part of the pilot for one-to-one laptops for the Houston ISD. The laptops enabled teachers to be proactive in planning instruction and the students to have access to instruction and the curriculum at any time of the day. For our students who needed to work, play sports, participate in fine arts, or look after their younger brothers and sisters, the laptops and the learning management system proved to be essential to students' success. Within our engineering and digital media strands, we upgraded the computers to the software requirements needed to provide the experiences and the instruction. Also, for both strands, the school purchased or was given the machines, tools, cameras, and other technology to complete the curricular requirements.

A second use of the technology allowed the teachers to implement the Universal Design for Learning (UDL) Framework. All students need different ways to access the curriculum, share their learning, and engage with the course and students even in STEM work. While some students can successfully learn utilizing the traditional classroom methods, many others need some assistance in one of those three components.

Finally, technology for the teachers cannot be underestimated or undervalued. STEM education must be supported by the core content courses. Both STEM and core content courses need to be able to track student data in academic and behavioral performance. Knowing a student's literacy and numeracy levels, state assessment scores, attendance rate, and other quantifiable variables can help with intervention and enrichment opportunities. When teachers take student data into planning, whether they use Excel, Google Sheets, Tableau, Mesa Ontime, or other analysis or visualization software, they will support more students along the way and are more likely to accomplish the outcomes of the shared vision.

As the CTE and STEM pathways matured, Chavez had fewer students wanting to drop out of the pathways. Several of the pathways also became extremely popular and eventually prolific in producing students who graduated with licenses and certificates. The Vet Tech program eventually produced nearly fifteen Vet Tech graduates per year. If the school had access to more veterinary offices, it would have easily produced a dozen certificates more a year.

However, this may not seem like a lot given the number of graduates. Keep in mind that Chavez is an urban school without many students who live in the country. The number of Vet Techs produced was among the highest in southeast Texas. The most popular program by far was the school's Pharmacy Technology program. Each year, it would grow by ten or twelve students, and each year, with a passing rate of over 95 percent, the graduates would pass the CCMA test and the Pharmacy Tech exams. Eventually, the school successfully licensed thirty-two Pharmacy Tech graduates in one year (the most for a high school in Texas). Meaning nearly one in eighteen graduates saw Pharmacy Tech as the way to pay for college or support themselves and their families. In the first graduating class, nine students of combined Chavez High School and San Jacinto College associate degree students graduated in 2019. Each of the nine who completed the program had a job offer before graduating. In 2021, the first graduating class of IBCP students graduated.

## Conclusion

In conclusion, I often think about the number of times that I would arrive at Chavez early in the morning to get a jump on the day's or week's work. I would park in the back and walk the length of the building to get to my office. The last area that I would walk through was the cavernous cafeteria. Even if I arrived before six in the morning, there was always at least one student in the cafeteria, either on their laptop or with their head down, sleeping. The scenario was repeated if I left the building after six in the evening. (The school day was 8:30 a.m.–4:00 p.m.)

Schools are often a place of refuge, heating, cooling, Wi-Fi, and electricity. Students, particularly those from historically marginalized populations, should see school as the place where the options for their and their families' futures are being developed. STEM education has a direct tie to providing for the families that our students will eventually have as parents, and it can also provide for the students' parents. Please take, for example, the collaborative Process Technology associate degree that Chavez and San Jacinto College developed; the starting salary for a student who graduates from high school and San Jac at the same time is over $75,000. With overtime, by the end of a graduate's first year, usually around when the student is nineteen or twenty years old, the students can earn a six-figure salary. The graduate can move into management with a bachelor's degree and earn even more. This promotion will not only change the graduate's life but also has the possibility of changing outcomes for the generations of the family that follow the graduate. My vision for STEM education is for schools and districts to see it as a catalyst for generational change for families and communities. To quote the school's namesake, Cesar E. Chavez, "Once social change begins, it cannot be reversed. You cannot un-educate the person who has learned to read. You cannot humiliate the person who feels pride. You cannot oppress the people who are not afraid anymore." STEM education will provide the social change needed to lift the students, the graduates, the school, and the community so that they can provide for themselves and others and take pride in what they have accomplished.

# About the Author

RENE SANCHEZ

Rene Sanchez serves as the new superintendent in the Champlain Valley School District in Vermont. Taking the role in Vermont is a return home for his family. His wife, Jean, is a native of Brattleboro and an alumna of the University of Vermont.

In South Bend, as the Assistant Superintendent for Operations for the South Bend Community School Corporation, Rene oversaw Human Resources and Transportation. He created the New Teacher training monthly series within HR, focusing on equity and Universal Design for Learning. Rene coordinated moving to a new HRIS system and digital personnel records. He spearheaded the purchase and implementation of placing WI-FI on the district's school buses in the transportation realm. This action led to South Bend Schools' bridging the digital divide during the COVID-19 pandemic and to recognition of the district in local, state, and national news.

As principal of Houston ISD's 3000+ student Chavez High School, his team founded the IB Diploma and Career programs, the Process Technology A.A.S program with San Jacinto College, was featured in the book, *UDL: Moving from Exploration to Integration* for its culture and use of Universal

Design for Learning schoolwide, and was recognized by America Achieves and the Organization for  Economic and Community Development for closing the academic gap with affluent schools. Scholarships increased from $6 million to over $19 million. And, in just three years, his vertical team increased their fine arts programs by 250 percent.

Originally from the Rio Grande Valley of Texas, Rene attended undergraduate studies at the University of Notre Dame. He earned a bachelor's degree in Government and International Studies. In addition, Rene has a law degree from Ohio State. From the University of Texas at Austin, and he has a master's degree in Education. He is currently enrolled at Indiana State to conclude his Ph.D.

He has been married to Jean for sixteen years, and they have three children and three dogs. He loves to cycle, cook, read, fish, golf, and work in the garden and yard.

# EIGHTEEN

# Encouraging Multiple Pathways to Graduation

## WITH AN EMPHASIS ON CAREER AND TECHNICAL EDUCATION

Cosimo Tangorra, Jr., Ed.D.

Multiple pathways to graduation, with an emphasis on CTE, should be encouraged in an effort to promote STEM skills and advance social justice. With a focus on the individual learning styles and interests of all students, public schools have never been better positioned to ensure that all students tap into their full potential. At the same time, an opportunity exists to address the current and future employment and economic needs of our nation. The provision of student-focused pathways to graduation that are future focused has the potential to guarantee that students leave public education with the skills that allow them to adapt to the changing economy they will encounter.

The issue of inequity has plagued public education in America for centuries. We have long known that our current system of educating young people has continued to work for some segments of the population but not others. Achievement gaps are a stain on our system that hold our young people back from future opportunities. When schools become culturally responsive and sensitive settings for children, all students will thrive, and society will benefit. We have to keep the work of equity and inclusivity and personalized learning practices at the forefront of our planning. Ensuring that all of our students have equal access to CTE pathways to graduation can serve to address the inequities that exist in our system and bolster the STEM economy.

Public education in the twenty-first century finds itself in an extremely turbulent situation. Yet, it is also well positioned to make inroads to equity and economic stability. An institution that was once venerated and looked upon as the great liberator is under attack; at the same time, it is positioned to demonstrate its significance and necessity. Both sides of the political spectrum are criticizing public education for what they believe are valid reasons. One side sees an extremely inefficient, and perhaps unnecessary, system that wastes valuable resources and no longer yields past returns. The other perceives an opportunity to bring about significant reform to ensure that the system no longer disenfranchises a substantial portion of the population it was designed to serve. This dynamic is occurring at a time when the institution is well positioned to ensure that all students are equitably served, and at the same time strengthen the nation's economy. A well-rounded academic program with an appropriate emphasis on CTE will allow for the creation of a system that stresses the individual learning characteristics and interests of the learner, while focusing on the current and future economic needs of the country.

In a 2013 Metropolitan Policy Program at the Brookings Institution report, Jonathan Rothwell stated:

Workers in STEM fields play a direct role in driving economic growth. Yet, because of how the STEM economy has been defined, policymakers have mainly focused on supporting workers with at least a bachelor's degree, overlooking a strong potential workforce of those with less than a BA.

The provision of an educational program focused on propping up the STEM economy has the potential to ensure that public education creates fulfilled, contributing citizens and can allow us to focus on the individual aspirations and abilities of all the students we serve. This goal can be obtained by engaging in redesign of public education as outlined in AASA's 2021 report, *An American Imperative: A New Vision for Public Schools*. Such a redesign can result in an emphasis on multiple pathways to graduation that stress lifelong learning, employability, personal interest, and life enrichment.

As we begin the process of reimagining education in light of COVID-19, we must ask ourselves the question: Why would we even consider returning to policies and practices that have proven to be inequitable and inefficient? We have an opportunity to address many wrongs and injustices that have existed for far too long.

As schools consider operating under a "new normal," our actions need to be responsive, compassionate, inclusive, nuanced, and clear. We can reimagine education, and we can make positive change from this experience, but we have to include all stakeholders in the process. Students, families, faculty and staff, higher education, the business sector, and the community at large must come together and rally around creating educational programs for K–12 students that adequately prepare them for the challenges that the future has in store for them.

Students not only need the foundational skills that have always been crucial but also need to have the freedom to pursue their desires and focus on their strengths. Families and caregivers need to be informed of the possibilities that exist for their children and the multiple pathways that exist to actualize their pursuits.

The push to ensure that all students would go to college forced schools to limit offerings and left many unfulfilled. In underserved communities, this push only exasperated the inequities that existed. The conditions that existed in these districts did not lend themselves to a focus on multiple pathways to graduation to begin with. Far too many offered few electives beyond the state requirements. I recall a district where the valedictorian was unable to gain acceptance into several state colleges due to the lack of depth in her transcript.

More often than not, students were engaged in college-bound pathways that did not connect the world they were experiencing or the economy that awaited them. Countless students and families incurred massive debt in pursuit of a college degree that would not lead to gainful employment. Due to an inability or unwillingness to provide other meaningful pathways, they were unable to develop the technical skills that were needed to fill the employment vacancies that existed in their communities. Concomitantly, employers were unable to recruit a workforce with the skills they needed, creating an avoidable crisis.

Throughout the course of my career, I have been concerned about the well-being of all of my students. As we know, the circumstances that surround certain students give us reason to be more concerned about some more than others. We all have suffered many sleepless nights fretting over students who are dealing with the unfair hand that life has dealt them.

Having spent the first two thirds of my career working in high-need districts, I had great concerns about my students who were suffering from the effects of poverty. Who wasn't going to have enough food over a weekend, over a break, over the summer? Who was not going to be properly supervised? Who had to witness the abuse of a loved one the night before or was abused themselves? The ravages of poverty can wreak havoc on the lives of students, causing learning to be the least of their worries.

However, I met the student that I was most concerned about when I began working in a very affluent suburban school district. Far from being a victim of poverty, he came from a privileged family and had more than any of the students I had worked with in the past. This child was an exceptional student who excelled in all of his coursework. At the time he was in his senior year and was taking seven Advanced Placement courses. He sacrificed a lunch period to ensure that he was taking all of the classes that he "had to take" in order to be successful. He came to me distressed over the grading policy of one of his teachers. The student was concerned that his grade in the class would prevent him from gaining acceptance in a particular pre-med program at a particular college.

At first, I was convinced he was simply overworked and was quite frankly blowing the issue out of proportion. Several conversations with him led me to become increasingly concerned about his state of mental wellness. Over time, I began to worry that this bright young man would take his own life if he did not get into the college of his choice. He was convinced that not doing so would ruin his chances of a fulfilling life. I took the appropriate steps with his family and with his school counselor to make sure he received the help he needed, but this experience, coupled with many conversations I had with other students and family members, caused me to look more closely at the unrealistic expectations of this community and the culture of the school district in general.

What I discovered was troubling in several aspects. It became clear that what was valued was a single pathway to graduation. One focused on heavy participation in Advanced Placement courses and acceptance to prestigious institutes of higher education. In fact, this particular pathway was the only one celebrated and recognized. Pathways with a focus on CTE or those that led to the military were looked down upon.

This placed undue stress on students, families, and faculty. Students that had interests and goals that would have led them to a different path were made to feel less than. Many students who had aspirations that would lead to the military or a technical career felt forced to engage in a pathway that was not appropriate for them. As a result, effective pathways were not easily accessible to all students, denying them the opportunity to learn the skills and competencies to engage in a fulfilling and productive school experience.

This equity dilemma led to the creation of a committee of stakeholders who would convene and craft a plan that would guarantee that all students have access to an appropriate pathway to graduation. The mission of the committee was to:

ensure that multiple pathways are made easily accessible and that these pathways effectively offer all students the opportunity to learn the skills, competencies and emotional intelligence to gain, keep and advance in a fulfilling and highly-productive career journey.

Ensuring that all students have equal access to all of the opportunities that a school district has to offer requires that we focus on each individual student. A single pathway to graduation that can be defined as better than any other should not exist. It cannot be said that one pathway is superior to another. For example, the pathway that includes multiple AP courses and acceptance to a prestigious institution of higher education has no greater value than a career and technical pathway that leads to technical school or directly into the workplace. The right pathway is determined by the student and that individual student's interests, goals, strengths, and aspirations. All pathways to graduation have equal value. The best pathway is determined by the student, in conjunction with caregivers, supported by school counselors, and based on the individual student's strengths and aspirations.

Our Pathways Committee initiated conversations with students, families, faculty, and alumni in an attempt to educate the community about the benefits of STEM and CTE education. We quickly learned the value of making students and families aware of the opportunities of STEM and CTE pathways as early as possible when, during a conversation with high school students, one senior expressed that she wished she knew that CTE opportunities were an option: "I would have made very different choices and had a much happier high school experience." This statement had an impactful effect on every committee member. The decision was made to focus on students and families in grades four through eight, as well as those in high school.

We also discovered that in addition to educating students and families, a considerable amount of professional development was needed for the faculty, especially for the school counselors. Even more important than the PD was the message from the Board of Education and the Leadership Team that all pathways to graduation were equally valued.

To help the community's perception of success evolve and to promote access to more CTE and STEM education, a rebranding of multiple pathways was necessary. The committee was charged with eliminating any stigma that was associated with alternate pathways and celebrating them. One method for celebrating a career and technical pathway was the strengthening of our relation with our local Board of Cooperative Education Service (BOCES). We highlighted the pathways that existed through our BOCES and its Center for Advanced Technology (CAT). These events were hosted at the CAT and highlighted not only the pathways that currently existed but also the potential careers and opportunities that would become available to students pursuing these other options.

We invited guests from local industry to explain their current labor shortages and to discuss the skills they were looking for in potential employees. Employers described how CTE pathways could lead to gainful employment that would solidly land them in middle class. Successful graduates from those pathways told stories of their ability to be engaged in fulfilling work while saving money for college or even having college paid for by their employer. Others were able to tell inquiring students and families of their ability to earn a middle class living without the burden of college debt.

STEM education is an integral component of every pathway to graduation. Traditional pathways focusing on STEM have created fulfilled, productive graduates for years in almost every imaginable field. However, countless other pathways to graduation are available to all of the students we serve. Career and technical pathways are all too often overlooked and undervalued, despite the fact that they can lead to equally fulfilling and enriching results. Regardless of the pathway any student selects, the process for that decision has to be inclusive and focused on the individual.

# About the Author

## COSIMO TANGORRA, JR., ED.D.

Dr. Cosimo Tangorra, Jr., superintendent of schools for the New Hartford Central School District, has extensive and varied management experience in P–12 Education. He has been a teacher, academic administrator, principal, and superintendent of schools. He also served as Deputy Commissioner of Education for the State of New York. School District Reorganization is an area of professional interest for Dr. Tangorra and was the topic of his doctoral dissertation.

An advocate for equity, he is dedicated to strategic planning that focuses on programs that prepare students for "their futures and not our past," modernizing school facilities, and community partnerships. In his chapter, Dr. Tangorra outlines how one school district worked to create a culture that embraces multiple pathways to graduation with an emphasis on the individual goals, strengths, and aspirations of all students.

Dr. Tangorra received his associate's degree from Herkimer County Community College, a bachelor of arts from Siena College, his master's degree from the College of Saint Rose, Certificate of Advanced Study from SUNY Cortland, and a doctorate in education (Ed.D.) from the Sage Colleges.

# NINETEEN

# Actions Carry More Weight than Words in STEM Education

## Dr. Johnnie Thomas

I will be the first to tell you how important STEM education is, but talking about it is not enough. We need to provide our students with the resources they need, not only to get quality jobs that allow them to provide for their families but also to grow up to become productive members of our community. Our society needs more scientists, engineers, and mathematicians to solve the problems of our ever-changing world and getting more workers in those fields starts with educating our students now.

Many of our students—90 percent of whom are Black and between 6 and 7 percent Latinx—grow up never meeting anyone who started a small business, ran a corporation, or invented a product. This opportunity and experience gap has profound implications on students' social and economic mobility later in life.

At Rich Township District 227, we address those gaps in a variety of ways, including setting high expectations for our 2,600 students. Our team has rallied around the Rich Township mission, which is to sustain a focus on students and on student success. Each day, we expect them to be prepared, engaged, and focused. They rarely fail us. We have also reimagined our curriculum to offer students true learning with a purpose: we merged the district's three high schools to create one "super school" with multiple campuses. A student's home base is determined by the career pathway he or she chooses as a first-year student—either science, technology, engineering, and math (STEM) or fine arts, business, International Baccalaureate (IB), and communications. Some kids must move between campuses for their classes; we provide shuttles, but it is up to them to build travel time into their schedules. They must set challenging learning goals for themselves, whether they are heading to college or the workforce.

Both embody our overall strategy of supporting student agency. We expect our students to take responsibility for their own learning. Agency means discovering what motivates them, striving for continuous improvement and rebounding from their "failures." It is not only the key to learning but the cornerstone of a successful life.

## Why STEM Education Is So Important

Employers are constantly on the lookout for viable employees with real-world experience in STEM fields, and we have millions of students graduating high school each year who could fill those empty positions if they are given the right opportunities in their education. Manufacturers and construction companies are at the top of the list of employers in need of qualified workers, and they need people who have had an education in one or more of the STEM fields. When we give our students access to STEM education, we provide them with opportunities, not only to be fiscally successful and provide for their families but also to build our cars, roads, bridges, buildings, and more.

Students with access to STEM educational materials and training have the added benefit of gaining stackable credentials. This means they can enter the workforce as soon as they graduate high school, or they have the option to pursue a college degree and build on the skills they developed in high school, leading to even more opportunities for their careers. What matters is giving them the choice to build the careers and lives they want for themselves.

When students are given access to a STEM education and can work in STEM fields, which tend to provide stable, lucrative jobs, this gives those students the opportunity to lift themselves out of poverty (or keep themselves and their families out of poverty). Families that are fiscally sound are less likely to rely on government programs and more likely to contribute to their communities and society, both fiscally and through volunteer programs.

## The Importance of Working Together

The subtitle of this book says it all: "It Takes a Village to Raise a 21st Century Graduate." I firmly believe that we are stronger together, and it is only by working together that we will accomplish our goals, both as educators and as a society. Just like we need more than one engineer and more than one scientist to provide the resources our society needs to grow and thrive, it takes more than one person to provide our students with the resources they need to grow holistically.

It is also important to remember that, as educators, we set an example for our students. When they see us working together to achieve a common goal, they will grow up with both an understanding of the importance of teamwork and the skills necessary to work in a team environment, which will better equip them to thrive in both their personal and professional lives. Teamwork is important in many industries, but it is especially important in STEM industries. Many manufacturers and construction companies require professionals from two or more of the STEM fields to work together to complete a project and to keep everyone involved safe throughout the duration of the project.

When students are raised to work well with others, they grow up to be better equipped to work on a team. That, in turn, will lead to better outcomes for the projects they work on.

## Transforming Education to Emphasize STEM

Preparing our students for careers in STEM fields requires much more than just a few tweaks to our educational system. In some cases, it is going to require major transformations, so let's look at some of the things we can do to provide our students with the resources they need to be successful STEM professionals.

### Relate STEM to Their Daily Lives

Students who sit in class thinking, *When am I ever going to use this?* are much less likely to engage with the material, much less pursue the subject outside of class or in subsequent classes. The key to engaging students is to show them how the material they are learning is relevant in their day-to-day lives.

Understanding how their phones work or how the foods they eat can impact their athletic performance are lessons that are much more likely to grab their attention than a list of facts and numbers.

*Training*

Education provides students with the knowledge they need but not necessarily the skillset for the jobs manufacturers and construction companies need to fill. Most employers want workers with real-world experience in STEM fields, which means we need to devote some time each semester to getting kids out of the classroom and into environments where they can put into practice the laws and theories they learned in the classroom.

The great thing about implementing this part of the educational transformation is that there are so many opportunities for making this happen. It could be as simple as a class project to design and build something or run an experiment. Or it could be something more involved, such as an apprenticeship program in which students can work alongside professionals in the field on projects that will have a direct impact on their communities. That has the added benefit of giving students an opportunity to experience what an average day in their chosen profession would be like, which better enables them to choose a profession.

*Mentorship Programs*

Mentorship programs are incredibly important for giving students opportunities to see what is possible for them and to start charting a path to get to a successful career, whatever that means for them. There is a reason our country's biggest companies invest in mentorship programs and encourage their employees to participate in them—it's because they understand that employees who have access to a mentor are much more likely to be successful in their jobs.

For students, the effect is amplified because access to STEM professionals and business leaders can significantly widen a student's perspective when it comes to understanding what their professional options are. This is one area where our educational system has consistently failed our students, especially in schools that primarily serve minorities. Schools whose students are most likely to live in poverty or belong to a racial or ethnic minority are least likely to have access to mentorship programs, and yet are most in need of such programs. When these students are introduced to STEM professionals and business leaders, that gives them the opportunity to learn how they can follow in the footsteps of those professionals, which makes them much more likely to follow that path out of poverty and lift their communities up with them.

*Start Teaching STEM Early*

We start asking kids what they want to be when they grow up almost as soon as they start talking, yet we don't give them access to all their options until much later in life, if ever. Many students don't get any introduction into the STEM fields until high school or even college, but if no one taught them to cultivate an interest in STEM prior to that point, they are much less likely to be engaged in that kind of learning. This is especially true if they had to wait until college before getting an introduction into the STEM fields, because by that point most students have already chosen their profession and are focused on taking classes that will help them get a job in that field.

Contrary to the widely held belief that STEM is only for advanced students, research has shown that kids can (and should) be introduced to the STEM fields as early as elementary school. Most young children are very curious about their world and how things work and giving them the answers they seek is a great way to introduce them to STEM fields. This leads back to my first point about making STEM relevant to students' daily lives to get them to consider STEM fields. The earlier you can accomplish that goal, the more likely they will be to pursue and maintain an interest in STEM.

## Conclusion

Educators have a lot of progress to make when it comes to implementing STEM into our educational system, and there is no way to make all the improvements we need to make at the same time. We will need the cooperation of educators at every level, as well as parents and guardians.

It is important to note that there is no end goal. Even if we think we know what the perfect STEM program looks like, that vision will likely change as we continue to see advancements made in all those fields, as well as in the educational system itself. But if we take even small steps toward filling those gaps and continue making progress toward our goal of a full STEM education for every student in America, we can change the future of this country.

# About the Author

## DR. JOHNNIE THOMAS

Dr. Johnnie Thomas is superintendent of Rich Township School District 227 where he has successfully brought innovation and transformation since 2017. The District 227 High Schools are unified in purpose with two campuses: the STEM Campus and the Fine Arts and Communications Campus. Under his leadership, District 227 focuses on the needs of the students while working with parents, staff, and the community. All students have Chromebooks and texts are digital. All students take part in AP courses and the IB program is strong. The district is financially stable with a balanced budget. GPAs are improved and more students are eligible for prestigious scholarships.

As former superintendent of Community High School District 155, Dr. Thomas was responsible for the successful delivery of comprehensive educational services to nearly 6,300 students who attend four high schools and an alternative education campus located in McHenry and Lake Counties. He is the eleventh superintendent in the district's ninety-five-year history and its first African American leader. Dr. Thomas' ascent within public school systems began with his role as a School Social Worker with Chicago Public Schools District 299. He was Executive Director for Student Services for

Valley View District 365-U from 2005 to 2009. He was Superintendent at Township High School District 214 where he led collaborative efforts to develop, implement, and evaluate services for special education students and faculty.

Dr. Thomas, a native of Chicago, received a doctorate from DePaul University, a master's degree from Loyola University, and his bachelor's degree from Southern Illinois University. He is an advisory board member of the Loyola University Chicago School of Education. He is a member of the Council for Exceptional Children, American Association of School Administrators, and Association for Supervision and Curriculum Development.

Dr. Thomas is a sought-after presenter and professional development facilitator on the topics of leadership, equity, differentiated education, and crisis intervention. He is the proud father of three children. Dr. Thomas is also a world traveler and particularly enjoys talking to global educators about how they prepare students to succeed.

TWENTY

# Expanding Opportunities Through Collective Impact

## LEVERAGING COMMUNITY PARTNERSHIPS
### Tamara Willis, Ph.D.

The Susquehanna Township School District is an urban fringe district located in central Pennsylvania. According to Niche.com, serving 2,990 students and representing 25 languages, the district is one of the most diverse school districts in the nation and the most diverse district in the Commonwealth. With learners and families representing an expanse of experiences, ethnicities, religions, socio-economic statuses, and post-secondary ambitions, our diversity is our greatest strength and our biggest challenge.

Like a growing number of school districts across the nation, we struggle to raise test scores, address unfunded mandates, and thrive despite dwindling resources. With a deluge of competing priorities, school leaders risk becoming preoccupied with local urgencies at the expense of preparing learners to meet the challenges of a technologically transformed workplace, leaving us wondering how to prepare our increasingly diverse student populations to excel in the Information Age.

Our district is no different. Just seven years ago, we engaged a cross-section of stakeholders to develop our mission and vision. These guiding tenets were not a laundry list of hollow aspirations awaiting the occasional review by school leaders. They were reminders of our agreement and commitment to learners and families entrusting us with their futures. Our vision is *World-class, every day, in every way*, and our mission is *the success of every learner*. We are acutely aware that, to fulfill our vision and mission, we must maintain a global perspective on education and the impact that the educational experiences we provide will have on every learner's success.

Developing our core values further solidified the district's commitment to providing STEM equity and access through a collective approach.

- Every learner has the right to a world-class education.
- We will be a leader in innovation and technology.
- The learning environment must be safe and supportive.
- Our diversity is our strength.
- Community partnerships are vital to our success.
- We will act ethically at all levels of the organization.

Districts may enjoy some level of individual impact through programmatic and curricular adjustments. However, collective impact positions districts and their partners to mutually identify strategic challenges, foster innovation, demonstrate organizational agility, and achieve a shared mission and vision on behalf of *every* learner. Pooling resources across sectors provide comprehensive and sustainable STEM experiences that extend beyond the school walls. For example, STEM instruction at local libraries, before and after school programs facilitated by service organizations, and summer programs operated through the local municipality further enhance the community's STEM ecosystem.

We began our journey by defining our district's STEM ecosystem, school, home, and community-based resources that, when aligned, would provide a wide range of learning opportunities for students. Further, we developed a list of partners representing public and private sectors with whom we conducted business. Finally, we identified additional partners who might share the district's vision for a STEM-focused community. Next, we developed the district's mission statement for STEM. It was critical to have a variety of perspectives in the room to ensure that our mission and vision reflected the expanse of our experiences. This work was accomplished over two days, offsite, with a cross-section of district staff. After this process, we developed our rules of engagement. Our district's mission and vision would drive its long-term commitment to improve all post-secondary outcomes. Our STEM mission statement was the catalyst to engage partners in a STEM-centered collective.

In their article, *Collective Impact*, John Kania and Maria Kramer (2011) identified collective impact as a promising approach to address large-scale social challenges. Within the framework, Kania and Kramer note five conditions essential in achieving sustainable change.

- Backbone organization
- Common agenda
- Shared measurement
- Mutually reinforcing activities
- Continuous communication

## Backbone Organization

The backbone organization offers structure to the process and monitors the overall progress of the initiative. School districts are uniquely positioned to serve as backbone organizations, particularly in developing a STEM collective. As hubs of their respective communities, districts deliver access to learners and typically hold transactional relationships with local business and industry leaders as a cost of doing business. School leaders who regularly leverage their influence within the community are ideal to launch a collective impact approach. Leaders who have yet to explore resources within their respective communities are strongly encouraged to build their professional cross-sector network.

## Common Agenda

The care and attention expended in developing the district's vision should be applied when advancing a common STEM agenda, as it will guide the work of partnership organizations. Essential to the development of a common agenda is the purposeful identification of businesses and organizations within the public and private sectors that share an interest in strengthening the local economy, extending STEM-related opportunities for all learners, and establishing local and regional STEM partnerships through workforce development. One conversation at a time, school leaders must invest time to meet with leaders across sectors to share their vision for the district, the learners, and the community's future. A combination of passion, vision, and trend data will help identify allies whose needs and capacities complement the district's mission and vision for connecting learners to STEM-related careers.

The road to a common agenda is unquestionably time-consuming, given the growing responsibilities of school leaders. However, it will pay dividends for all stakeholders when carefully crafted. Despite mounting demands, this endeavor is paramount to achieving systems-level change. Budgetary constraints are a mainstay in education. While fiscal support can certainly enhance program opportunities, collective impact also necessitates a commitment of human resources, professional expertise, and preservice experiences, not easily monetized contributions.

Finally, a shared agenda, particularly one aimed at expanding STEM learning outcomes for underserved and underrepresented student groups, must be viewed through an equity lens. Identifying policy and procedural changes needed to address systemic inequities will also facilitate the removal of barriers that reinforce disparities within STEM disciplines.

The needs of my district are always at the forefront of my mind as I engage in conversations. I ask, how can this partnership support our mission of the success of every learner? How can I make my mission their passion? How can I align our vision with their area of expertise? The leader's role is to make the connection and make it meaningful for my would-be partners.

## Shared Measurement

Mutual indicators of success are as critical as a common agenda. Partners need to envision the fruits that collective impact will yield for their organization. Key questions to shape the development of shared measures include, what does a successful partnership look like for each stakeholder? Will your district serve as a pipeline for partnering organizations' workforce? Will partners be allowed to mentor and shape potential employees through internships, shadowing, and other preservice experiences? What is a reasonable timeline, and how will you benchmark progress along the way? While the agenda will guide the work, shared measurements ensure the quality and degree of the work accomplished. Each partner must articulate outcomes they hope to realize and a timeline for achieving those outcomes.

## Mutually Reinforcing Activities

Collective impact, though akin to collaboration, is distinct in its focus on the individual accomplishments of each partnering organization. Following creating a common agenda and shared measures of success, each organization agrees to align its efforts to pursue the collective vision. For example, partners whose policies prohibit high school students from job-shadowing or internships commit to reviewing and updating those policies to extend such opportunities to our learners. Similarly, organizations with limited experience connecting with K–12 institutions commit to developing processes to advance authentic and productive partnerships.

Learning requires greater customization to meet learner needs, and the prevalence of information at each learner's disposal weakens the argument that schools exist to impart knowledge to generations of learners. Instead, schools must help learners channel the proliferation of information through moral, ethical, and creative lenses that will ultimately lead to a college/career pathway. Because authentic STEM experiences at key intervals across the K–12 experience are critical in building comprehensive pathways and ensuring access for all learners, our district is committed to developing new and advanced

STEM courses beginning in middle school. Counseling learners to consider career and technical education (CTE) courses in high school and partner with post-secondary institutions to offer scholarships and dual enrollment opportunities into STEM majors. Partner organizations created opportunities to engage high school learners through internships, mentorships, job-shadowing, and targeted funding.

## Continuous Communication

Ongoing communication is vital to collective impact as partner organizations revisit the agenda to determine if it remains appropriate and ensures that activities remain aligned. Continuous focused communication with partners fosters trust and allows the coalition to monitor progress toward achieving shared measures and pivot where warranted. Partners are also more readily available to troubleshoot, solicit, and offer feedback on shared challenges when communicating regularly.

## The Leader's Charge

Districts must equip and produce authentic STEM experiences in cutting-edge and emerging technologies to prepare learners to compete within the rapidly evolving and increasingly competitive marketplace. Businesses and industries seek a workforce that can innovate, think critically, and generate the next big idea, and preparation cannot begin in high school. On the contrary, STEM experiences must be integrated throughout the K–12 experience and beyond the school walls. Even the highest performing school districts will have limited impact if assuming this challenge alone. Harnessing resources beyond the district to include business, industry, and civic organizations can intensify the district's impact to reach *all* learners. As chief visionaries and advocates of the organization, it is incumbent upon school leaders to assist would-be partners in seeing their role as impactful within the STEM ecosystem and then guide the larger coalition to achieve those shared goals. Every handshake, every phone call, every conversation is a strategic move in building that coalition. Do not miss an opportunity to capitalize on the moment. Students need you.

## Reference

Kania, J., & Kramer, M. (2011). Collective Impact. *Stanford Social Innovation Review, 9* (1), 36–41. https://doi.org/10.48558/5900-KN19

# About the Author

## DR. TAMARA WILLIS

Dr. Tamara Willis is unwavering in her campaign to prepare globally competitive, 21st-century learners through innovation and customization. In her chapter she reveals how school leaders can expand access to STEM opportunities for all learners using a collective impact approach. The focus of a Microsoft Customer Story and two-time featured speaker for HP's *American Reinventors Series*, Tamara has successfully leveraged key partnerships to develop sustainable STEM initiatives within her school district. During her twenty-two years in education, she has served as a building-level principal, an assistant superintendent for curriculum and instruction, and an adjunct professor in the areas of special education and instructional leadership. Dr. Willis is currently Superintendent of the Susquehanna Township School District located in south central Pennsylvania.

# TWENTY-ONE

# Setting the Stage for STEM Success

## ESTABLISH AN EFFICACIOUS ENVIRONMENT

### Dr. Deborah L. Wortham

As education pioneer John Dewey neared death, he summarized what he had experienced in his field. According to Dewey, "What I have learned is that the purpose of education is to allow each individual to come into full possession of his or her personal power" (cited in Sarason, 1993, p. 78). Educators must provide the environment for students to achieve the success that generates from commitment and effort. Educators must manage the development of students so students may acquire the self-confidence that will enable them to have a quality life.

As a teacher and a superintendent, I've often heard teachers engage in conversations that determined which students would have a quality life and which would not. A teacher said, "These poor inner-city children just can't learn, and there's nothing I can do to help them."

> Research indicates that such comments may be more than idle chatter; they may reflect important beliefs that influence teacher-student interactions and teachers' success in producing student achievement gains. Researchers have labeled these teachers' attitudes as their sense of efficacy—the extent to which teachers believe they can affect student learnings. (Dembo & Gibson, 1985, p. 173)

The research harnessed my pledge to instill in teachers the importance of believing that all children can learn and that teachers can bring about positive student change. It required a paradigm shift. As Sarason (1993) stated, "Those who advocate for an educational reform seek not to reform themselves but to change someone or something else" (p. 19).

As teachers and administrators, how efficacious are we? We educators must ask ourselves several questions:

- Do we use that power?
- Do we have positive expectations?
- Do we provide social-emotional support for our students?
- Or do we deny that power and accept poor performance as characteristic of the environment and pardon the students?

The answers to these questions can control the way staff teaches and administrators lead. A wealth of literature substantiates teacher efficacy as a framework for student achievement. In a study of 240 teachers, Podell and Soodak

(1993) revealed that "teachers who perceive themselves as inef-fectual consider regular education inappropriate for under-achieving students from low socio-economic status (SES) families; teachers who believe that they are effective do not differentiate students by SES" (p. 247). These perspectives are the blueprint used to establish policies and procedures for students, schools, and school systems.

The philosophy of the Efficacy Institute (a nonprofit organiza-tion in Lexington, Massachusetts) espouses a belief system based on the premise that all children can learn to the highest level (Howard, 1995). This belief system is transferred to a pedagogical framework or teaching practice that involves all educators in the educational development of young people.

The instructional environment calls for the infusion of efficacy principles into the existing curriculum throughout the school day. Through teaching practices such as efficacy, we can make a difference in the academic and social-emotional behaviors of the students. The philosophy of efficacy is the basis of creating and maintaining a culture of confidence and for organizing the commitment and investment to build a quality education system for all children, and it enables students to explore interest in careers in Science, Technology, Engineering, and Mathematics (STEM).

However, "Our approach to educating children is failing because the attitudes that underlie it are wrong" (Howard, 1987, p. 6). The crux of the problem resides in what we think about people and their capacity to learn; we operate on the premise that people are born smart or not smart. A self-fulfilling prophecy then negates the ability of all children to learn and to learn all subjects. Consider an administrator or teacher who does not believe in their ability to educate students against all odds. Consider students who do not believe they can succeed.

"Efficacy is the capacity to mobilize available resources to solve problems and promote development" (Howard, 1987, p. 5). Efficacy is a process that defines our approach to children and learning. Howard (1993) defined development as "the process of building identity, character, analytic and operational ability, and self-confidence. Development is the basis of the individual's capacity first to understand and contribute to the community's goals and then to envision and pursue new ones" (p. 4). In essence, it is the framework to excel in STEM.

After an introduction to the philosophy of efficacy, I read the literature about teacher efficacy, participated in professional development, and visited the forty-two Efficacy Project schools.

Too often, disadvantaged children succumb to negative influences and negative self-images. Too often, society perpetuates those negative expectations. The problem is that some teachers, staff, students, and parents do not have a sense of efficacy. They do not believe that all students can learn, and they do not believe they have what it takes to make a difference in their students' lives. Yet, research proves they can make a difference.

I intended to have all teachers believe that they could make a difference in student achievement and possess a sense of efficacy. Because teachers are the change agents, particular attention should be paid to the way teachers instruct and the thoughts affecting their actions. Because administrators are the key to instructional improvement, specific attention should be paid to the decisions made concerning instruction. Are these decisions based on the philosophy of efficacy? Are these decisions based on the District's mission, vision, values, and goals?

Every principal and teacher is accountable for increasing student achievement. The critical question is, how? How do you increase student achievement for all subgroups? Many will agree that the key is through establishing an efficacious environment.

An efficacious environment is more than memorializing a series of affirmations. The real key to school improvement is more than a series of affirmations and "feel good" statements but an actual bonding of staff, building confidence, signaling effort (not zip code) as the key to achievement, regular assessment, and reflection upon student learning. Most importantly, the key to school improvement is to believe in the value of an efficacious learning community, having a plan to sustain it, and being willing to overcome obstacles.

As an administrator, I aimed to improve student achievement in a high-performing and a low-performing high school by implementing and adhering to the school's mission statement based on the belief that all children can learn. The mission statement was used as a framework to improve student achievement. Anecdotal and biographical data were maintained and analyzed. The results point to at least five declarations that can be made, all of which highlight the capacity principals and superintendents have. Principals and superintendents can: (1) create a positive culture; (2) foster positive relationships; (3) create personal and professional development, (4) facilitate the change process, and (5) incorporate all voices.

In January of a particular school year, the Superintendent of Schools summoned me to her office. I was serving as the Director of Professional Development. What could I have done this time? Why did she want to talk? Was the discussion going to be positive, negative, or somewhere between the two? The superintendent explained that I would need to take on an additional role: principal of a high school. Imagine serving in both capacities. Why me? I was a former principal of a high-performing elementary school and a K–8 school, which convinced the superintendent that I could continue in the role of District administrator and simultaneously lead the 1,200 student body high school. The high school was already designated a Blue Ribbon High School and, in academics, ranked in the top 10 in the state. The challenges with the school rested in its leadership not focusing on the school's mission. The long-term tenure of the previous principal ended abruptly as a new principal was hired. For unknown reasons, the new principal began isolating himself, did not engage staff or students, and did not address, reinforce, articulate, or live the school's mission. Staff, students, and the community called for his resignation between August and January. I indeed accepted the challenge and, in fact, started the new role the next day.

On the first day of the principalship, I introduced myself by walking the halls and visiting classrooms. No one knew about the reassignment of the former principal. Just one week prior, students were refusing to go to class. They were in an uproar because the pep rally was canceled due to poor behavior: roaming the halls, cutting class, leaving campus during closed lunch periods, using drugs, and so forth. Teachers and administrators enforced consequences, hoping to curb the negative behaviors, by calling the city police, spraying students with mace, and giving them the ultimatum to go to class or go home. Parents were outraged by the consequences enforced by the teachers and administrators. This is what I inherited. What could I do? I chose to shift the environment and encourage students to return to class by merely asking them, "Where are you going to college?" They responded with the name of a college (true or not). My response was, "You can't get there from here [the hallway]!" This choice and its outcomes are a lesson in the value, power, and capacity for positive change that principals have when they choose to embrace an efficacious environment. Because of the overwhelmingly positive response by students, the first professional development session with teachers was devoted to reviewing the mission statement. Does it respond to the three questions?

1. What do we want to do?

2. How will we know if we are achieving it?

3. What will we do to guarantee success?

In that session, the commitment was reengineered! They rallied around the reason the school existed. The school existed (according to the mission statement) to get students into college. They rallied around the rich history of the school and the strong alumni. They rallied around the real responses to the three questions and provided data to prove that they were achieving their goal. They spent time discussing ways to guarantee success. They knew that they could not continue in the manner in which they were headed. They had lost sight of their mission. They needed someone to navigate the course again. They left the session with a determination to work collaboratively and with the new principal and stay focused on its mission and commitment to an efficacious environment. They attended Efficacy seminars. They each were assigned a senior student to mentor. The goal was to work with each senior on AP and STEM courses and support their quest for acceptance into college.

At the end of the school year, they had the highest graduation rate of all forty-three high schools in the District! The students received a total of $17,000,000.00 in college scholarships and financial aid. Out of the 350 graduates, two enrolled in Armed Services, and the rest in a two-year or four-year college or University! Oh, let's not forget the lone summer school graduate!

The second experience mirrored the previous. However, now I'm a superintendent in another District. The message of success at the Blue Ribbon High School (that regressed) was put to the test at a high school that was not a magnet school, which had never met state standards and was in Corrective Action. Could the same strategies that led to the restoration of a magnet school be used to lead another school forward with totally different demographics? I understood the importance of the role of the mission statement, creating an efficacious environment, and having staff "live" the statement and participate in Efficacy seminars. The process of creating the mission statement mirrored the previous District. Stakeholders came together to discuss, "Why do we exist?" The statement responded to the three questions:

1. What do we want to do?

2. How will we know if we are achieving it?

3. What will we do to guarantee success?

In addition to state requirements, students were requested to take the SAT and apply for college. Teachers were assigned as mentors to support students as they adhered to the requirements. A Commitment Ceremony was held to toast and commemorate the relationship between the senior, the mentor, and the parent. Logs were maintained to document support given to the senior by the mentor. One student asked, "Why do I have to take the SAT and apply to a college? No one in my family ever graduated from high school or attended college. Why should I?" This student later stated, "I am going on a college tour! I am going to college!"

Adhering to the school's mission will enable you to accomplish your goal of improving student achievement. At the graduation for this class, eighty-five of the ninety-two seniors met the state and local requirements. Five of the students would graduate from summer school, and two would become fifth-year seniors. All eighty-five graduates received acceptance letters into a two- or four-year college! This represents an unduplicated count of fifty-eight colleges or universities. Scholarships totaled over $640,000.00 (does not include financial aid)! It is true—leadership is leadership! Living the mission statement based on belief in efficacy is a sure-fire way to improve student achievement and promote success in STEM.

In one of the lowest-performing District's in the state of New York, it happened again! This time in a middle school. The staff committed to embracing the philosophy of efficacy; they participated in Efficacy seminars and implemented a new STEM curriculum. Additionally, a new wing was built on the school to accommodate the increased enrollment (due to the STEM theme), science and technology labs, and curriculum. Ultimately, in 2018, all twelve K–8 schools in the District were identified by the New York State Education Department as "Good Standing!" One school was designated "International Baccalaureate."

*A leader knows the way, goes the way, and shows the way.*
*~John Maxwell*

# References

Dembo, M., & Gibson, S. (1985). Teachers' sense of efficacy: An important factor in school improvement. *Elementary School Journal* 86(2), 174–184.

Howard, J. (1993). The third movement developing Black children for the 21st century. The state of Black America. New York: National Urban League.

Howard, J.(1987). Getting smart: The social construction of intelligence. The efficacy seminar for educators. Lexington: MA1 The Efficacy Institute.

Podell, D., & Soodak, L. (1993). Teacher efficacy and bias in special education referrals. *Journal of Educational Research* 86(4), 247–253.

Sarason, S.B. (1993). *The Case for Change*. San Francisco: Jossey-Bass.

# About the Author

DR. DEBORAH L. WORTHAM

Dr. Deborah L. Wortham is an educator, a professor, an author, an elder, and nationally recognized on the lecture circuit. A native of Chicago, Illinois, Dr. Wortham is prepared spiritually and educationally. She was the first locally elected Superintendent of the Roosevelt Union Free School District in eleven years. During her tenure, the District moved from "State Takeover" to "Good Standing." She is the former Superintendent of the School District of the City of York in York, Pennsylvania, and Steelton-Highspire School District in Steelton, Pennsylvania. Dr. Wortham served as the first African American female Superintendent of the East Ramapo Central School District in Spring Valley, New York, where all twelve of the K–8 schools moved to "Good Standing." Still fond of the Roosevelt Union Free School District, currently Dr. Wortham serves as the Superintendent in charge of "Resetting Roosevelt!"

Her credentials include: a doctorate in Educational Leadership from Nova Southeastern University, Fort Lauderdale, Florida; a master's degree in Reading from Morgan State University, Baltimore, Maryland; and a bachelor's degree in Elementary Education from the University of Wisconsin, Madison, Wisconsin.

Her most recent community highlights include 2020 Top 100 Most Influential Educators in New York State; the Martin Luther King, Jr. Service Award; Rockland County School Board Association Award of Excellence; Mid-Hudson School Study Council, Educational Leadership Award; First Timothy Christian Church for Outstanding Contributions; and Leadership Award, Nassau County, New York NAACP. Additionally, she is the recipient of the following awards: Washington-Rose "Inspiring Woman" Award; Service Above Self Award, Central Nassau County of Rotary International; Salute to Unsung Sheroes' of Sisters in the Struggle Award, Nassau County; and Educator of the Year, Long Island Black Educator's Association.

Dr. Wortham and her husband, Rev. Chester H. Wortham, have led marriage and leadership development seminars at various conferences. They have two children, Dr. Shelley and Min. Chester, III, one son-in-law, Sheldon, one daughter-in-law, Marquita, one granddaughter, Jordynn Lynn, and four grandsons, Chester, IV, Sheldon, Jr., Zachary, and Ethan.

To God be the Glory for the Great Things He has Done!

# Afterword

**21ˢᵗCentEd.**

The United States is on the cusp of a STEM crisis. The problem? Data[1] indicate that not enough of our nation's students are equipped for STEM opportunities now, much less in the future. The performance of American teenagers in reading and math has stagnated since 2000.[2] US students' academic achievement lags behind many of their peers in other countries.[3] Our children are being taught 20th-century skills using 19th-century pedagogy for a rapidly accelerating 21st-century global economy. It doesn't add up to a winning formula. When our teachers lack access to comprehensive STEM resources, they simply aren't equipped to prepare students for future success.

The good news is that there is a solution. By providing multiple STEM experiences to ALL learners starting at an early age in school, after school, and out of school, **you** are doing your part to solve the STEM crisis. We must deliberately include girls, students of poverty, students of color, rural students, and urban students. 21stCentEd works with school districts, charter schools, private/parochial schools, homeschools, parents, and community organizations to provide a comprehensive STEM education in this way.

## Benefits of working with 21stCentEd

1. A Virtual STEM Academy for ALL students with unlimited in and out-of-school access to self-paced, project-based, courses organized through career pathways. Early STEM engagement brings strong benefits to students at a crucial time when many young learners may have formed or are forming self-defeating ideas about their abilities.

2. Each enrollment provides access to over 900 hours of STEM in 46 (and counting) courses, 20 hours of instruction with 10 projects per course. Students are

exposed to more than coding and robotics, i.e., *Design Thinking, Entrepreneurship, Financial Literacy, Digital Media and Art, Animation, Coding, Gaming, Robotics, Engineering, Virtual Reality, Artificial Intelligence, Machine Learning, Internet of Things (IOT), and more, including various certifications.*

3. A built-in, self-paced curriculum, with lesson plans to address STEM-teacher shortages. Mentors and coaches (who can be anyone: teachers, community partners, and/or students, etc.) who guide students through courses, in school, after-school, and out of school.

4. Establish or enrich current CTE programs with courses in *Architecture and Construction, Audio/Visu al Technology, Health Science, Transportation, Engineering and Technology, and much more.*

5. Collaborate with community partners such as libraries, Boys and Girls Club, YMCA, community centers, etc. to provide access beyond the walls of the schools. *There simply isn't enough time in the day for schools to do the work alone. It truly takes a village to raise a 21st-century graduate.*

## The Problem

One of the greatest challenges facing the United States is motivating, educating, and training a technical workforce capable of competing on a global scale. Nationwide, there are over 665,000 open computing jobs currently available.[4]

By 2028, 2.5 million[5] advanced manufacturing jobs will go unfilled, and 2.7 million[6] data science and analytics jobs will be open. The success of our children and economy is at stake.

*Research from STEM Connector[7,8] reports five STEM talent gaps:*

Fundamental Skills Gap

- The average STEM competency is insufficient for modern work.
- The new "STEM skills" (employability and 21st-century skills, including *communication, collaboration, critical thinking, and creativity*) are hard to build in traditional education models.

- Employers do not always clearly or appropriately define the skills they need.

## Belief Gap

- Students hold incorrect beliefs about their own STEM ability.
- Students and others feel they do not "belong" in STEM.
- Students falsely believe only certain industries offer STEM jobs.
- Employers hold incorrect beliefs about how academic performance and credentials relate to job success.

## Postsecondary Education Gap

- Not enough people hold credentials beyond high school, but most STEM jobs require postsecondary credentials.
- Credentials are misaligned with employer needs.
- Jobseekers and employees face challenges engaging in lifelong learning.

## Geographic Gap

- In some regions, the number of jobs is declining, leaving people out of work and with outdated skillsets.
- In some regions, there is a booming STEM economy, but companies must import talent as locals are not equipped to compete for those jobs.

## Demographic Gap

- Lack of access to resources drives achievement gaps in STEM education from early years.
- Bias and historic inequity remain embedded in education and employment systems.

The STEM pipeline, which is meant to guide students from elementary school into successful careers, has not been effective. The skills gap is *real*.

## The Solution: Comprehensive STEM

While transforming our society to match the pace of technology is a heavy lift, it is a necessary one. We work with school districts, charter schools, private/parochial schools, homeschools, parents, and community organizations to close the STEM talent gaps by focusing on the following areas:

- Access to high-quality STEM learning opportunities.
- Representation of girls, racial and ethnic minority groups, and persons with disabilities in STEM.
- Essential 21st-century knowledge and skills,

including critical thinking, creativity, collaboration and communication.

- Stakeholder collaboration in education, workforce development and economic growth.

## Increasing Access

By launching a virtual STEM Academy with 21[st]CentEd, schools of all shapes and sizes will be able to provide STEM access in-school, after-school, and out-of-school. Students will have access to the kind of learning that will help them achieve mastery-level education in STEM skills.

Courses are flexible by design, suited to your school's individual needs and budget concerns. Part self-guided, part mentor-led, students get access to the expert-level knowledge that's just not attainable in an 8-hour school day guided by standardized test results.

The first step in solving this problem is to create and enhance STEM experiences for ALL students from an early age. This means instilling a STEM culture throughout the community. We must deliberately include girls, students of color, and both rural and urban students in our STEM initiatives.

Data[9] show that while representation in STEM has slowly risen, the proportion of racial and ethnic minority groups in science and engineering remains below their share of the college-educated workforce.

Cultivating the United States' 21st-century workforce demands ALL students be provided multiple STEM experiences early in their development and throughout K–12. To do this, we need a comprehensive STEM plan—and advocates who are ready to support it.

When learners don't have access to STEM resources in class, after-school programs, and out of school, they are missing out on valuable skill-building time.

The 21stCentEd STEM platform allows charter schools, private and parochial schools, school districts, after school organizations, and individual homes to launch a Virtual STEM Academy to create an environment in which ALL students have STEM experiences from an early age. We assist school districts by implementing a comprehensive STEM initiative that reaches students anywhere, anytime.

## Meeting Students Where They Are

To make STEM curriculum relatable, students need a chance to explore the various skills in each competency. 21stCentEd's virtual academy is available digitally, making our STEM resources accessible to students, no matter where they live. Enrolling in 21stCentEd's STEM academy provides multiple course options in the areas of:

- *Design Thinking, Entrepreneurship, Financial Literacy, and Game Development.*
- *Digital Media, Graphic Design, Digital Art, and Animation.*
- *Programming and Coding, Robotics, and Engineering.*
- *Virtual Reality, Artificial Intelligence, Machine Learning, and the Internet of Things (IOT).*
- *And more, including various certifications.*

Working through different STEM-based courses can become a catalyst for a student's future career. By giving students access to STEM resources early and often, we're preparing them for the future.

## Why It Matters

*Yes, You CAN Learn STEM.*

The way to increase the number of students studying STEM is to provide ALL students with STEM experiences early in their development. Equitable access to comprehensive STEM resources ensures every learner has an opportunity to explore the opportunities available to them.

The learners most at risk of being left behind are students of poverty, urban and rural youth, girls, and students of color. Our program immerses ALL students in project-based STEM experiences before they form self-defeating ideas about their abilities—*they will experience STEM-related successes before they think they can't.*

## Sparking Interest

Our approach introduces students to STEM through fun, engaging courses that build multiple competencies. Because the 21$^{st}$CentEd online platform has built-in curriculum and lesson plans, it eliminates traditional barriers to launching a comprehensive program. All of our courses are self-paced to prevent students from becoming overwhelmed. And, anyone can be a STEM mentor or coach with 21$^{st}$CentEd—the curriculum doesn't require a trained STEM educator, of which there is a shortage,[10] to drive success.

By piquing interest in STEM careers early, your community members, businesses and other stakeholders will reap economic benefits.

## STEM Now Equals Success Later

When students are immersed in STEM-based experiences from an early age, they're less likely to develop self-defeating ideas about their abilities

later. When a school district invests in STEM-based education, they are investing in workforce development, economic viability and most importantly, the future success of our children.

Remember, we are already feeling the impact of the U.S. STEM shortage. Over 665,000

STEM-related jobs are sitting unfilled. If we don't do something to fix it now, what will that number look like by 2030?

## Who Has Been Successful

The second-most ethnically-diverse school district in the commonwealth of Pennsylvania, Susquehanna Township School District, serves nearly 2,900 students. After realizing the dire need for STEM workers, superintendent Dr. Tamara Willis selected 21ˢᵗCentEd to jumpstart her district's STEM initiative.

"The research behind 21ˢᵗCentEd is riveting," said Dr. Willis. "To hear Marlon and the rest of the 21ˢᵗCentEd team share current numbers regarding the STEM workforce was alarming. I knew we needed to do something."

Dr. Willis collaborated with county commissioners, the school board and the technology director to implement 21ˢᵗCentEd courses at the elementary, middle and high school levels. "We discussed the current state of the workforce and the existing challenges in trying to find qualified workers for STEM-related jobs," Dr. Willis said. "As a group, we've committed to acting as STEM advocates for our community."

Students have ample time to access 21ˢᵗCentEd's STEM courses. In the district's elementary building, stu- dents can access courses through the school's library media specialist. Middle schoolers leverage 21ˢᵗCent- Ed courses during project-based learning time, as well as at the district's after-school STEM club.

"The benefit of 21ˢᵗCentEd's curriculum is the ability to give every student several opportunities to explore STEM, which opens the door to career pathways they may not have explored otherwise," Dr. Willis said. "We're removing barriers to help our learners succeed, and parents are pleased that our district is actively offering high-level opportunities to explore STEM."

21ˢᵗCentEd courses provide exposure and exploratory opportunities to students, according to Dr. Willis. "The fact that 21ˢᵗCentEd courses are challenging, but not graded, is a huge selling point for our students," she said. Students can dive into

STEM lessons without the lingering fear of receiving a poor grade.

The STEM crisis *is real*.

"A problem of this magnitude may leave district leaders wondering where to begin," said Dr. Willis. "The fact is, investing in 21$^{st}$CentEd as a STEM partner is a strong starting point."

If we don't address the shortage of STEM workers now, we are likely to experience an economic crisis and jeopardize the United States' position as a leading innovator in STEM industries.

## Launching a STEM Revolution

As education leaders, it is *our* responsibility to provide opportunities for STEM learning to every learner.

An effective STEM initiative takes time. The transformation of a community, town or region into a vibrant STEM community is a three- to five-year project to start – anything less can result in wasted resources, frustration, loss of credibility and demotivation.

However, the benefits provided to students, communities and the country are well worth the investment.

## Getting Started

The need to change the way we educate our stu dents is clearer than ever. This is true for all kinds of educational models.

So whether you're a school district, a charter or magnet school, or a private or parochial school, providing students with access to additional STEM learning resources is crucial. If you are a school superintendent, a school principal, an executive director or a headmaster, the pressure is on to get it right. Thankfully, you can't go wrong with 21$^{st}$CentEd.

The first step is to provide access to comprehensive STEM resources to ALL learners.

*Be the STEM advocate your students need. Our future—and our children's future—depends on it.* **Let's get them ready!**

# Notes

## Introduction

1. Sloan, Willona M. 2012. *What Is the Purpose of Education?* ASCD. https://www.ascd.org/el/articles/what-is-the-purpose-of-education.
2. PDK Poll. 2019. *Frustration in the Schools: Teachers Speak Out on Pay, Funding, and Feeling Valued.* September. https://pdkpoll.org/wp-content/uploads/2020/05/pdkpoll51-2019.pdf.
3. Frey, Carl Benedikt, and Michael Osborne. 2013. *The Future of Employment: How Susceptible are Jobs to Computerisation?* September 17. https://www.oxfordmartin.ox.ac.uk/downloads/academic/future-of-employment.pdf.
4. Manyika, James, Susan Lund, Michael Chui, Jacques Bughin, Jonathan Woetzel, Parul Batra, Ryan Ko, and Saurabh Sanghvi. 2017. *Jobs Lost, Jobs Gained: What the Future of Work Will Mean for Jobs, Skills, and Wages.* November 28. https://www.mckinsey.com/featured-insights/future-of-work/jobs-lost-jobs-gained-what-the-future-of-work-will-mean-for-jobs-skills-and-wages.
5. World Economic Forum. 2020. *Jobs of Tomorrow Mapping Opportunity in the New Economy.* January. https://www3.weforum.org/docs/WEF_Jobs_of_Tomorrow_2020.pdf.
6. Toffler, Alvin. 1970. *Future Shock.* New York: Bantam Books.
7. The Conversation. 2020. *Computer Science Jobs Pay Well and are Growing Fast. Why are They Out of Reach for So Many of America's Students?* December 9. https://theconversation.com/computer-science-jobs-pay-well-and-are-growing-fast-why-are-they-out-of-reach-for-so-many-of-americas-students-150419.
8. Deloitte. 2018. 2018 Skills Gap in Manufacturing Study Future of Manufacturing: The Jobs Are Here, but Where Are the People? https://www2.deloitte.com/us/en/pages/manufacturing/articles/future-of-manufacturing-skills-gap-study.html.
9. Callum Rivett. 2020. PwC Report Forecasts 2.7m New Data Science and Analytic Jobs Will Be Created by 2020. May 17. https://technology-magazine.com/data-and-data-analytics/pwc-report-forecasts-27m-new-data-science-and-analytic-jobs-will-be-created-2020.
10. Morgan, Steve. 2021. Cybersecurity Jobs Report: 3.5 Million Openings in 2025. Cybercrime Magazine. November 9. https://cybersecurityven-

tures.com/jobs/.

11. Ibid.

12. 21stCentEd. https://21stcented.com/.

# 1. The Need for STEM with Latino Students

1. Student Research Foundation. 2020. *Hispanics & STEM: Hispanics are Underrepresented in STEM Today, but Gen Z's Interest Can Change the Future.* https://www.studentresearchfoundation.org/wp-content/uploads/2020/04/Hispanics_STEM_Report_Final-1.pdf.

# 2. STEM Programs versus a STEM Education

1. Business Wire. 2020. *Global IT Outsourcing Market 2020–2025 – Growing Demand for Efficiency and Scalable IT Infrastructure – ResearchAndMarkets.com.* August 10. https://www.businesswire.com/news/home/20200810005303/en/Global-IT-Outsourcing-Market-2020-2025---Growing-Demand-for-Efficiency-and-Scalable-IT-Infrastructure---ResearchAndMarkets.com.

2. National Science Foundation. 2021. Women, Minorities, and Persons with Disabilities in Science and Engineering. https://ncses.nsf.gov/pubs/nsf21321.

# 11. Elementary STEM Building Blocks

1. Does your school have a guaranteed and viable curriculum? McREL.org, https://www.mcrel.org/does-your-school-have-a-guaranteed-and-viable-curriculum/

2. Next Generation Science Standards, https://www.nextgenscience.org/

3. Standards for Mathematical Practice, http://www.corestandards.org/Math/Practice/

4. National Science Teaching Association, https://www.nsta.org/bestsellers

5. TechTrep™, https://www.techtrep.com/

6. A Plan 4 Excellence: Valley Stream UFSD Thirteen Strategic Plan 2016–2021, Goal 1

7. NYSED Smart Start Grant, http://www.nysed.gov/edtech/smart-start-grant-program

# Afterword

1.  STEM Education in the U.S. (2017). https://www.act.org/content/dam/act/unsecured/documents/STEM/2017/STEM-Education-in-the-US-2017.pdf
2.  Programme For International Student Assessment [PISA] Results (2018). https://www.oecd.org/pisa/publications/PISA2018_CN_USA.pdf
3.  Pew Research Center, U.S. Students' Academic Achievement (2017). https://www.pewresearch.org/fact-tank/2017/02/15/u-s-students-internationally-math-science/
4.  Why Computer Science? https://code.org/promote
5.  BLS Data, Oxford Model, Deloitte Manufacturing Skills research initiative.
6.  BHEF, Investing in America's data science and analytics talent (2017). https://www.pwc.com/us/en/publications/assets/investing-in-america-s-dsa-talent-bhef-and-pwc.pdf
7.  STEMConnector, State of STEM. https://www.stemconnector.com/state-stem-report/
8.  STEMConnector, Input to Impact: A Framework for Measuring Success Across the STEM Ecosystem (2019). https://www.stemconnector.com/wp-content/uploads/2019/06/STEM_Report_FInal_2019-2.pdf
9.  The State of U.S. Science and Engineering (2020). https://ncses.nsf.gov/pubs/nsb20201/u-s-s-e-workforce#women-and-underrepresented-minorities
10. Essentials of Social Innovation (2011). https://ssir.org/articles/entry/collective_impact; Facts to Know about Teacher Shortages (2018). https://www.nctq.org/dmsView/Teacher_Shortage_Fact_Sheet

Made in United States
North Haven, CT
09 February 2022

15890990R00147